AF622065

Diseases and Management of Field, Medicinal and Fodder Crops

NIPA® GENX ELECTRONIC RESOURCES & SOLUTIONS P. LTD.
New Delhi-110 034

About the Author

Dr (Mrs) Gayatri Biswal (1965) passed B. Sc (Ag), M.Sc (Ag) and Ph.D degree from Odisha University of Agriculture and Technology, Bhubaneswar, Odisha. Awarded Gold Medal in M. Sc (Ag) in Plant Pathology. She has published 100 research papers in international and national journals, eight book chapters. She has guided one Ph.D student, 15 PG students in Plant Pathology and 20 MBA students in Agri Business Management. She is co-guide of three Ph.D students of Nematology and also 25 PG students of Entomology and Nematology of Odisha University of Agriculture and Technology, Bhubaneswar. She is the life member of seven professional societies. She is Fellow of Indian Phytopathological Society (2018), Fellow of Hind Agri-Horticultural Society (2018), Fellow of Applied Zoological Research Association (2022)

Diseases and Management of Field, Medicinal and Fodder Crops

Gayatri Biswal.Ph. D. FPSI , FHAS, FAZRA
Department of Plant Pathology
College of Agriculture
Odisha University of Agriculture and Technology
Bhubaneswar, Odisha, India

NIPA® GENX ELECTRONIC RESOURCES & SOLUTIONS P. LTD.
New Delhi-110 034

NIPA® GENX ELECTRONIC
RESOURCES & SOLUTIONS P. LTD.

101,103, Vikas Surya Plaza, CU Block
L.S.C. Market, Pitam Pura, New Delhi-110 034
Ph : +91-11-43860225, Mob.: +91 9717133558, 9540816132
E-mail: newindiapublishingagency@gmail.com
Website: www.nipaersources.com

© 2026, Author

Print ISBN: 978-93-58870-91-6

ebook ISBN: 978-93-58870-92-3

All rights reserved. No part of this publication may be reproduced, stored in a retrieval system or transmitted in any form or by any means, including electronic, mechanical, photocopying recording or otherwise without the prior written permission of the publisher or the copyright holder.

This book contains information obtained from authentic and highly reliable sources. Reasonable efforts have been made to publish reliable data and information, but the author/s, editor/s and publisher cannot assume responsibility for the validity, accuracy or completeness of all materials or information published herein or the consequences of their use. The work is published with the understanding that the publisher and author/s are not attempting to render any professional services. The author/s, editor/s and publisher have attempted to trace and acknowledge the copyright holders of all material reproduced in this publication and apologize to copyright holders if permission and/or acknowledgements to publish in this form have not been taken. If any copyrighted material has not been acknowledged, please write to us and let us know so that we may rectify the error, in subsequent reprints.

Trademark Notice: NIPA®, the NIPA® logos and their presentations (the way they are written/presented) in this book are the trademarks of the publisher and hence may not be used without written permission, if copied or used without authorization, the infringer will be prosecuted as per law.

NIPA® also publishes books in a variety of electronic formats. Some content that appears in print may not be available in electronic books, and vice versa

Composed and Designed by NIPA®.

Dedicated to
My Beloved Parents
Smt. Labanya Biswal
Late Sri Narendra Nath Biswal
and My Loving Daughter
Late Bhagyashree Khuntia (Bulbul)

Preface

The book "Diseases and Management of Field, Medicinal and Fodder Crops" is a little attempt to fulfill the needs of students and teachers. It is the compilation of class lectures by the author for post-graduate students with required information on diseases of field and medicinal crops. The descriptions are brief, informative with reference to recent work in India. Mistakes and omissions are expected to be found in the text. Hence teachers and students are requested to put forth their valuable comments and necessary suggestions for further improvements of the book.

In writing this book the literature on the diseases of field and medicinal crops developed by various agencies and organizations like ICRISAT, IRRI has been freely used. I extend my sincere thanks to the authors and editors of various books, journals and periodicals which has been used as reference materials in this book. Every care has been taken to cite the bibliographic references. However, any omissions, misrepresentations, incorrect citations or other mistakes that have occurred are regretted.

I am very much grateful to all my teachers and colleagues in the Department of Plant Pathology, College of Agriculture, Odisha University of Agriculture & Technology, Bhubaneswar for their help in various ways during the preparation of this text book. I express my gratitude to the NIPA, Genx Electronic Resources & Solutions P. Ltd., New Delhi for fulfilling my dream in bringing out the book timely and nicely.

I am thankful to my husband Mr. Bhagirathi Khuntia , my two sons Mahima Prasad, Bibhudutta and brothers Dillip, Pradeep, Pratip and Sandip for their support and encouragement.

I am very much thankful to all my friends, relatives and brother and sister in-laws, nephews and nieces. Special thanks to my students and farmer friends who are the source of inspiration to me throughout my academic journey.

Odisha University of Agriculture & Technology
Bhubaneswar

Dr (Mrs) Gayatri Biswal

Contents

1

Diseases of Cereals

RICE (*Oryza sativa*)

Fungal diseases

1. Blast

Symptoms: All above ground parts (leaf, leaf sheath, culm) of the plant are affected. There are three type of blasts (**leaf blast, node blast, neck blast**)

Leaf blast: Eye or spindle shaped spots on leaves (margin is dark brown and centre becomes ashy) and leaf sheath are observed. Several spots appear on the leaves cause blightening.

Node blast: In severe cases culms and nodes are affected. Darker lesions are observed.

Neck blast: If the disease attack after panicle emergence the neck (the stem portion below the panicle or base of the panicle) is affected. Before grain filling if the disease occurs the earhead becomes chaffy. If the disease attacks after grain formation the ear head may break or hang down due to weight of the grains which results in maximum damage.

Causal organism: *Pyricularia grisea*

Management

i) Burning of straw and stubbles should be practised or infected crop debris should be removed and deeply buried in soil.

ii) Deep summer ploughing

iii) Removal of collateral weed hosts

iv) Split application of nitrogenous fertilizers

v) Use of certified seeds or seeds from disease free crop can be used.

vi) Seed treatment with Carbendazim (2gm/kg seed) or Tricyclazole @ 0.1%

vii) Some varieties like Co.4,Co.25,Co.26, TKM1, ADT25, Tetep, Tadukan and Peta are highly resistant against blast which have been used as parents in breeding programme.

viii) Growing of tolerant varieties like IR-8, IR-20, IR-24, Jaya, Ratna and Vijay.

ix) Some regional tolerant varieties which can be grown are Lalat, Gouri, IR-36, Pratikhya and Swarna

x) Spraying of Tricyclazole 75% WP (0.1%)/ Carbendazim 50% WP @ 0.15% is recommended after appearance of the disease in the field. If necessary two to three sprayings may be practised. In endemic areas one spray should be given at the nursery, two spray during tillering stage and one or two spraying during panicle emergence stage should be followed.

xi) Antibiotics like Blasticidin-S @20 ppm and Kasugamycin 3% SL @ 2.5 ml/litre can be used as spraying.

2. Brown spot

Symptoms: All aerial parts (leaf, leaf sheath, stem and grains) of the plant are affected. Round or oval brown spots appear on leaves and leaf sheaths. In severe cases these spots may coalesce and cover the entire leaf. The infected leaves may dry out. Brown to black velvety lesions appear on glumes. Grains may be discoloured

Causal organism: *Helminthosporium oryzae*

Management

i) Burning of straw and stubbles Infected crop debris should be removed and buried in deep soil.

ii) Seed treatment with Vitavax Power or Carbendazim 50% WP (2gm/kg seed)

iii) BAM 10, Co.20, SR 26B, Padma and IR-24 varieties are resistant to disease.

iv) Growing of regional tolerant varieties: Lalat, Gouri, IR-36, Parijat, Pratikhya, Swarna

v) Spraying of Carbendazim 12% + Mancozeb 63% WP (0.1%) is recommended after appearance of the disease. If necessary two to three sprayings may be practiced at 7-10 days interval.

3. Sheath blight

Symptoms: Greenish-grey lesions with irregular dark brown margin on culm/ leaf sheath on the water level are noticed. At first the spots are about 1 cm long. They enlarge and may reach 2 or 3 cm in length and become greyish white with brown margin and irregular in outline. Later the lesions turn straw coloured and surrounded by bluish grey narrow bands. The leaf blade dries up from the tip downwards. Greyish black sclerotia are formed on lesions. Under humid conditions,the fungal mycelium spreads to the other leaf sheath and blades. Gradually the whole sheath rots and the affected leaf can be easily pulled off from the plant. In severe infection all the leaves of the plant are blighted resulting death of the plant.

Causal organism: *Rhizoctonia solani*

Management

i) Burning of straw and stubbles

ii) Seed treatment with Carbendazim 50%WP @ (2gm/kg seed)

iii) Growing of tolerant varieties: Pankaj, CR44-1, Krishna

iv) The disease is usually found in young plants under excess water-logged conditions. In that situation draining of fields will reduce the disease to a certain extent.

v) In severe infection or in disease prone areas soil drenching with Edifenphos and Kitazin (17% granules) during tillering stage is effective.

vi) Spraying of Hexaconazole 5% EC (0.2%) or Hexaconazole 5 EC 90.2%) or Propiconazole 25 %EC (0.1%) or Validamycin 3% L (0.2%) two to three times at 7 days interval is recommended.

4. Sheath rot

Symptoms: Panicles can not emerge fully. Irregular lesions having greyish centre with brownish margin are observed on upper most leaf sheath. The panicles are rotten showing aborted grains and powdery fungal growth inside the leaf sheath

Causal organism: *Sarocladium oryzae*

Management

i) Split application of N&K fertilizers

ii) Removal of collateral hosts

iii) Seed treatment with Vitavax Power (2gm/kg seed)

iv) Growing of tolerant varieties: Jayant,Triveni

v) Spraying of Hexaconazole 5% WG (0.2%)/Propiconazole 25% EC (0.1%) /Validamycin 3% L(0.2%)two to three times at 7 days interval is recommended.

5. Foot rot

Symptoms: The typical symptom is the excessive elongation of the stem that makes them several inches taller than healthy plants. In the nursery the affected seedlings are abnormally elongated become chlorotic and pale, thinner than the healthy ones. Such seedlings may die before transplanting. Some seedlings which survive the attack in nursery stage may die soon after transplanting. In the main field most characteristic symptom is the appearance of tall, stray, lanky tillers. Affected plants show infection at the collar region. The leaves show drying up starting from lower leaves. There is profuse branching of roots. Adventitious roots are also produced from first two to three nodes above the ground level. When such infected culm is split longitudinally and examined whitish cottony mycelium can be marked at the nodal regions. In severe infection pinkish fungal growth may be present on the base of the plant. Diseased plants produce few tillers. Generally such plants die before flowering but rarely ears are produced which are mostly sterile.

Causal organism: *Fusarium moniliforme*

Management

i) Summer ploughing, clean cultivation, removal or burning of straw and stubbles

ii) Seed treatment with Carbendazim 50%WP (2gm/kg seed)

iii) Growing of Co.18, Co.22, ADT.8, PTB.7 and GEB.24 which are highly resistant to the disease. Other regional tolerant varieties are IR-20, IR-32,IR-38.

iv) Basal drenching with Metalaxyl 8%+ Mancozeb 64% WP(0.2%) after appearance of the disease in the field. If necessary two or three times this chemical can be applied at 7 days interval.

6. Stem rot

Symptoms: The disease appears after transplanting in the form of small black irregular lesions at the water line on the leaf sheath. In favourable condition the discolouration spreads up and down from the water line. If such affected plants are cut open and examined the interior of the culm may be filled with greyish fungal mycelium. Sclerotia bodies are developed on base of the plant, leaves and leave sheaths. Numerous sclerotia may be found either loose inside the culm or embedded in the host tissues. The leaf blade dries up from the tip to downwards. Such infected stem rots and falls off. In the main field the affected plants collapse due to weakening of the stem base. In advanced stage the basal stem upto 2-3 inches turns dark green and also may rot. The spikelets remain chaffy and partially filled.

Causal organism: *Sclerotium oryzae*

Management

i) Summer ploughing, clean cultivation and burning of straw and stubbles

ii) Seed treatment with Carbendazim 50%WP (2gm/kg seed)

iii) Basumati 3, Basumati 370, Muskan 7 and Muskan 41 are resistant varieties in Punjab.

iv) Growing of tolerant varieties like Swarna, Pratikhya is recommended in Odisha condition.

v) Avoid standing of water in the field for longer period and in endemic area the drainage facility should be improved.

vi) Basal drenching with Metalaxyl 8%+ Mancozeb 64% WP (0.2%) twice or thrice after appearance of the disease in the main field can effectively check the disease.

7. False smut

Symptoms: The disease appears after flowering. The infected kernel is transformed into a large velvety yellow to orange masses which may be more than twice to thrice the diameter of the normal grain. Only a few grains in each panicle are affected. Infected grains are covered with spore masses which are disseminated by wind cause infection in healthy flowers. The disease usually occurs in the year of heavy rain fall or in bumper harvest season.

Causal organism: *Ustilaginoidea virens*

Management

i) Summer ploughing, clean cultivation, burning of straw and stubbles

ii) Use of certified seed

iii) Collection of diseased grains and buried them in soil or destroy them immediately. This will check the secondary spread of the disease.

iv) Spraying of Copper oxychloride (0.3%) in case of severe infection.

8) Narrow Brown Leaf Spot

Symptoms: Light brown lesions with dark centre are seen on the leaves and leaf sheaths. The number of lesions increases with the age of the plant. Due to heavy infection the leaves may dry.

Causal organism: *Cercospora oryzae*

Management

i) Burning of straw and stubbles

ii) Seed treatment with Carbendazim 50% WP (2gm/kg seed)

iii) Spraying of Carbendazim 12% + Mancozeb 63% WP (0.1%) is recommended after appearance of the disease. If necessary two to three sprayings may be practiced at 7-10 days interval

9) Stackburn

Symptoms: The spots on the leaves are pale brown with dark brown margin. The spots on the glumes are pale brown to whitish bounded by dark brown margin. It causes grain discolouration.

Causal organism: *Alternaria padwickii*

Management

i) Seed treatment with Carboxin 37.5% WP +Thiram 37.5% WP (2gm/kg seed)

ii) Hot water treatment of seeds at 54°C for 15 minutes after soaking the seeds in ordinary water for 12 hours has to be followed before sowing.

10). Bunt

Symptoms: The symptoms appear in mature grains only. Usually a few grains are affected in the ear. The disease can be detected as minute black streaks or pustules brusting through the glumes at the time of ripening. As the sori are hidden by glumes the smutted kernels inconspicuous in the field as well as in threshed grains unless the sorus is broken. In some cases whole grains are transformed into black sooty powders which may shed before harvest and can be seen adhering to awns and on adjacent leaves. The seedlings are weak and stunted when partially infected grains are used for sowing.

Causal organism: *Neovossia horrida*

Management

i) Seed treatment with Carbendazim 50% WP (2gm/kg seed)

ii) Growing of resistant varieties

11) Grain discolouration

Symptoms: The symptoms are visible on the grains. Dark brown to black lesions /spots appear on the grains. Sometimes full grain may be affected. There is reduction in the market value of the grains. The affected seeds show poor germination and vigour of the seedlings are affected.

Causal organism

i) *Curvularia* sp.

ii) *Drechslera* sp.

iii) *Fusarium* sp.

iv) *Xanthomonas* sp.

Management

i) Summer ploughing

ii) Clean cultivation

iii) Removal of weeds and stubbles

iv) Seed treatment with Carbendazim 50% WP (2gm/kg seed)

v) Spraying of Carbendazim 12% + Mancozeb 63% WP (0.1%) is recommended after appearance of the disease. If necessary two to three sprayins may be practiced at 7-10 days interval

12) Udbatta

Symptoms: A straight dirty coloured hard cylindrical spike considerably reduced in size emerge from leaf sheath much resembling like an incense stick/agarbati. Leaves of tender seedlings show greyish white appearance. No grains are formed on the affected ear.

Causal organism: *Ephelis oryzae*

Management

i) Use of certified seeds.

ii) Growing of resistant varieties

iii) Seeds of disease free field should be used for sowing.

iv) Hot water treatment of seeds should be done at 50-54^0 C for 10 minutes before sowing is very effective against this disease

v) Solar treatment of seeds is mostly sucessful in killing the pathogens present in the seed

vi) Seed treatment with Carboxin 37.5% WP +Thiram 37.5% WP (2gm/kg seed)

13) Leaf smut

Symptoms: The leaf smut disease is characterized by the appearance of distinct,linear,rectangular or angular black spots on the leaves. Their size is about 0.5-2.0 x 0.5-1.5 microns. These spots remain covered by epidermis which is ruptured when the leaves are soaked in water for some time and then the black smut spores are exposed visible.

Causal organism: *Entyloma oryzae*

Management

i) Clean cultivation

ii) Removal and destruction of infected crop debris.

iii) Deep summer ploughing twice after harvest of the crop.

iv) Use of resistant varieties

Bacterial diseases

1) Bacterial Leaf Blight

Symptoms: The disease appears in two phases

i) Kresek phase

ii) Leaf blight phase

i) Kresek phase: The leaves roll completely,droop and turn yellow or grey and ultimately the tillers wither away. Affected plants may be completely killed. This is the most destructive phase. It may be due to systemic infection.

ii) Leaf blight phase: Yellow or straw coloured stripes with wavy margins started developing from tip to downwards which may be observed on both sides. The leaves turn straw yellow. This phase usually occurs 4 to 6 weeks after transplanting. The grain formation may be affected in diseased plants. The grains may be partially filled or may be chaffy.

Causal organism: *Xanthomonas oryzae* pv. *oryzae*

Management

i) Summer ploughing, clean cultivation and removal of stubbles and weeds.

i) Use of certified seed

ii) Draining of standing water in the field from time to time. It should not pass through adjacent healthy fields.

iii) Alternate drying and flooding the field.

iv) Balance dose of NPK fertilizers is to be applied in the field with split dose of potashic fertilizers.

v) Growing of resistant varieties IR-20,IR-24, RP-4-14,RP-5-32, Pusa-2-21,Vijaya, Shakti.

vi) Hot water treatment of seed at 54°C for 15 minutes after soaking the seeds in ordinary water for 12 hours has to be followed before sowing.

vii) Draining of standing water in the field from time to time. It should not pass through adjacent healthy fields.

viii) Spray the crop with the mixture of Copper oxychloride (0.2%) and Streptocycline(0.01%) twice or thrice at 10 days interval.

2) Bacterial leaf streak

Symptoms: Appearance of interveinal long or short water-soaked greyish lines on leaves. In severe infection the entire leaf becomes yellow or dirty white.

Causal organism: *Xanthomonas oryzae* pv. *oryzicola*

Management

i) Use of certified seed or disease free seeds from reliable sources

ii) Growing of resistant varieties IR-20, IR-24, Bala,Sabarmati,Padma

iii) Hot water treatment of seed at 54°C for 15 minutes after soaking the seeds in ordinary water for 12 hours has to be followed before sowing.

iv) Spray the crop with the mixture of Copper oxychloride (0.2%) and bacterial Streptocycline (0.01%) twice or thrice at 7-10 days interval after appearance of the disease in the field.

Viral diseases

1) Tungro

Symptoms: The older leaves turn yellow orange starting from leaf tip and spreading downwards. The young leaves become slightly twisted and show mottled appearance. Plants become stunted which bear poor panicles with empty glumes. In highly susceptile varieties if the infection takes place at early stages the plants may die before flowering.

Causal organism: Rice tungro bacilli form virus(RTBV) & Rice tungro spherical virus(RTSV)

Vector: *Nephotettix virescens*

Management

i) Summer ploughing, clean cultivation, burning or removal of straw and stubbles

ii) Use of certified seed.

iii) Roguing out diseased plants from the field and should be buried in soil.

iv) Control the vector by spraying Fenthion 100 EC @ 400 ml/ha. In disease prone areas first spraying should be given 10 days after sowing. In the main field spraying should be given at 15,30,45 and 60 days after transplanting in the field.

2) Grassy stunt

Symptoms: Grassy appearance of the plants. Diseased plants remain stunted and show profuse tillering.

Causal organism: Grassy stunt virus

Vector: *Nilaparvata lugens*

Management

i) Summer ploughing, clean cultivation,burning of straw and stubbles

ii) Useof certified seed

iii) Control the vector by spraying Fipronil 5SC @2ml/litre

1) Orange leaf disease

Symptoms: The disease is characterized by an intense orange colour which may later change to golden or brownish yellow. In later stage there may be rapid rolling of leaves and rapid death of the plant.

Causal organism: Orange leaf virus

Vector: *Recilia dorsalis*

Management

i) Summer ploughing, clean cultivation, burning of straw and stubbles

ii) Use of certified seed

iii) Control the vector by spraying Fipronil 5SC @2ml/litre

2) Yellow dwarf

Symptoms: Chlorosis and yellowing of leaves. Plants become stunted and show profuse tillering. Affected plants die before reaching the maturity.

Causal organism: *Phytoplasma oryzae*

Management

i) Summer ploughing, clean cultivation, burning of straw and stubbles

ii) Use of certified seed

iii) Control of the vectors by spraying insecticides like Imidacloprid (0.25%) after appearance of the disease in the field. If necessary two or three spraying must be given at 10 days interval.

Nematode diseases

1) Ufra disease or Rice Stem Nematode

Symptoms: There are two types symptoms

i) Swollen ufra

ii) Ripe ufra

Swollen ufra: In this case the panicles remain enclosed within the leaf sheath.

Ripe ufra: The peduncles turn dark brown. The panicles emerge and produce normal grains only near the tip.

Causal organism: *Ditylenchus angustus*

Management

i) Summer ploughing

ii) Crop rotation with non-host crops.

2) Rice Root Nematode

Symptoms: Reddish discolouration of leaves and stagnation of plant growth. Brownish lesions on the roots and eventual rotting of root system with remarkable reduction in grain yield.

Causal organism: *Hirschmanniella oryzae*

Management

i) Field sanitation

ii) Balanced dose of fertilizer application

iii) Treatment of nursery beds with D-Dmixture@ 1litre/25 sq.mtr

iv) Transplanting of healthy nematode free seedlings

v) Before transplanting seedlings should be dipped in 500ppm solution of Carbofuran 12 hours before transplanting.

3) White tip nematode

Symptoms: Leaf tips turn white or pale yellow. Flag leaves shortened and twisted at their apical portions. Maturity of panicles is delayed

Causal organism: *Aphelenchoides besseyi*

Management

i) Spraying of infected crop with Trichlorophos (0.03%) three times if necessary has been found effective

ii) Hot water treatment of seed at 54°C for 15 minutes after soaking the seeds in ordinary water for 12 hours has to be followed before sowing

4) Rice Root Knot nematode

Symptoms: The nematode attacking the rice crop in nursery stage. It may also attack the upland rice. The leaf tips change to brown colour from the margin towards the mid rib. Height of the plants reduce In case of severe infestation club shaped galls formed on the roots of infected plants. Growth of root tip is retarded and profuse development of extra side roots are also observed

Causal organism: *Meloidogyne graminicola*

Management

i) Seed trearment

ii) Soaking of seeds with Carbofuran @500ppm solution. In case of transplanting condition root dipping is to be practiced, i.e. 12 hours dipping before transplanting.

Non-parasitic diseases

1) Tip burn of rice

Symptoms: The leaves loss the lustre become bronze coloured and later show scorching at the tip. The drying progress down along the margins and finally the whole leaf dries. Finally the entire plants dry and wither. Leaves and stalks become brown. Yellow and dried plants may be seen scattered throughout the field.

Causal organism: Physiological

Management

i) Green manuring

ii) Fields should be drained off at early stage of plant growth about 30 to 50 kg/ha of Ammonium sulphate should be applied.

2) Khaira disease

Symptoms: Leaves of diseased plants show chlorosis. brown to bronze coloured spots appear on the lamina surface. As the disease advances these spots increase in size and coalesce to form bigger spots and ultimately the whole leaf turns bronze coloured and dries. Diseased plants become stunted. Main roots turn brown and the finer roots are damaged. Plants do not produce ears. In some cases ears may produce but number of grains are less.

Cause: Zinc deficiency

Management

i) Two sprays of mixture of 2kg of zinc sulphate and 1 kg of slaked lime in 400 litres of water per acre can be recommended

ii) First spraying is to applied after appearance of the disease

iii) Second spray 10 days after interval

iv) In endemic areas a spraying schedule is recommended, i.e. mixture of 2kg of zinc sulphate and 1 kg of slaked lime in 400 litres of water per acre after appearance of the disease in the field. Second spraying is applied 25 days after first spraying. If necessary third spraying is followed at 15-20 days after second spraying.

Other minor diseases

Fungal diseases

Sheath rot *Acrocylindrium oryzae*

Banded sclerotial disease: *Corticium sasakii*

Seed discolouration *Curvularia* sp

Seedling blight,foot rot and grain discolouration *Drechslera sorokiniana*

Seedling blight *Sclerotium rolfsii*

Leaf spot *Leptosphaerulina oryzae*

Phoma seedling blight *Phoma glomerata*

White leaf streak *Ramularia oryzae*

Root rot *Rhizoctonia bataticola*

Leaf scald *Rhynchosporium oryzae*

Downy mildew *Sclerospora macrospora*

Leaf and glume lesion *Ascochyta oryzae*

Stack burn *Trichoconis padwickii*

Bacterial diseases

Bacterial sheath rot *Pseudomonas oryzicola*

Leaf and glume lesions: *Pseudomonas syringae* pv. *panici*

Bacterial panicle blight: *Burkholderia glumae*

MAIZE (*Zea mays*)

Fungal diseases

1) Pythium Seed and Seedling blight disease

Symptoms: Seedlings become pale yellow or dry up just after germination. The seedlings can be easily pulled out. Rotting of seeds may be seen. The disease is common in high soil moisture condition during rainy season. Death of seedlings occur in patches. Stalks of the affected plant sometimes show rotting at first internode just above the soil level and become brown and soft. The affected plants lodge but immediately do not break as the vascular bundles at that time are not damaged.

Causal organism: *Pythium aphanidermatum*

Management

i) Field sanitation

ii) Growing of resistant varieties

iii) Seed treatment with Carboxin 37.5% WP +Thiram 37.5% WP (3gm/kg seed)

2) Black bundle

Symptoms: Blackening of vascular bundles which are seen as black dots on the cut ends of the stalk. In severe cases leaves dry and plants may wilt. Ears may not be formed if formed they may show rotting.

Causal organism: *Cephalosporium acremonium.*

Management

i) Field sanitation

ii) Seed treatment with Carboxin 37.5% WP +Thiram 37.5% WP (2gm/kg seed)

3) Wilt

Symptoms: The disease affects both young and adult plants. The plants show dry up in the early stage of the growth. At the time of cob formation stage drying of top leaves may be noticed. Drying started from leaf tips and progress downwards. Gradually the entire plant dries up. Cob may not be formed if formed the grain formation is partial. White to light pink mycelial growth may be seen in the pith.

Causal organism: *Fusarium moniliforme*

Management

i) Field sanitation

ii) Seed treatment with Carboxin 37.5% WP +Thiram 37.5% WP @3gm/kg seed.

iii) Crop rotation for atleast 3 years

iv) Growing of resistant varieties i.e. Ganga Safed-2 and Ganga-101

4) Smuts

a) Common smut

Symptoms: In this disease galls are produced in the ears, axillary buds,tassels, stalks rarely in the leaves. Galls are dull white or greyish in colour much bulged from the surface and shiny and later on become dark. They rupture to expose black powdery mass. When formed on the cob they cause extensive damage. The spores are also observed on the seed coat.

Causal organism: *Ustilago maydis*

Management

i) Growing of smut resistant varieties.

b) Head smut

Symptoms: The tassel is partly or wholly converted into smut spores. Galls are first covered with a membrane that soon breaks open and powdery mass of spores expose. If only parts of the tassel and ears are converted into galls and the floral bracts turn into leafy structures. Affected plants become reduced in height and show the appearance of phyllody. The disease is both external and internal seed borne.

Causal organism: *Sphacelotheca reiliana*

Management

i) Seed treatment with Carboxin 37.5% WP +Thiram 37.5% WP (2gm/kg seed).

ii) Follow up the crop rotations

iii) Field sanitation.

5) Turcicum Leaf blight

Symptoms: Light olivaceous to brown coloured lesions appear on the leaf blade and these may extend into leaf sheath. Finally the leaf become straw coloured. The lesions first appear on the lower leaves. In susceptible varieties all leaves are affected. It is a seed borne disease. It spreads by air.

Causal organism: -*Helminthosporium turcium*

Management

i) Tolerant maize varieties are P.E. M. H-5, Bibek-21, Bibek-23, Kanchan-2, Nityashree

ii) Spraying of Mancozeb 50% WS @ 1250gm/500 litres of water per hectare sprayed at 10 days interval at least 2 times. If necessary another one or two spray can be given.

6) Maydis Leaf blight

Symptoms: Reddish brown spots appear on the leaves. Finally the colour becomes reddish yellow with dark brown borders. These spots coalesce with each other causing blightening of leaves. It is a seed borne disease. It spreads by air.

Causal organism: *Helminthosporium maydis*

Management

i) Tolerant maize varieties are P.E.M.H-1,H.Q.P.M-1,H.Q.P.M-5, Shaktiman-33, Shaktiman-4

ii) Mancozeb 50% WS @1250gm/500 lit of water per hectare sprayed at 10 days interval at least 2 times. If necessary another one or two sprays can be given.

7) Rust

Symptoms: It appears when the plant is at flowering stage. All above ground parts of the plants are affected but the rust pustules appear abundantly on both the surface of leaves. They may coalesce cause drying of leaves.

Causal organism: *Puccinia sorghi*

Management

i) Tolerant varieties are Bundal, Sheetal, H. H. M-1, H. H. M-2, H. Q. P. M. -1, Nityashree

ii) Mancozeb 50% WS @1250gm 500 litre of water per hactre sprayed at 10 days interval at least 2 times. If necessary another one or two spray can be given

8) Downy mildew

Symptoms: The disease produces characteristic downy growth on the leaves followed by yellow discolouration,brown and necrosis of blade and stunting of the affected plant. The disease is characterized by the chlorotic yellowish stripes on the leaves and the stripes are delimited by veins. The leaves become curled and yellow. The plants are dwarfed.

Causal organism: *Sclerospora philippinensis*

Management

i) Seed treatment with Metalaxyl 8% + Mancozeb 64% WP@ 2.5 gm/kg of seeds

ii) Spraying with Metalaxyl 8%+ Mancozeb 64%WP @2. 5 gm/litre of water.

iii) Growing of tolerant hybrids like Ganga Safed-2,Ganga-101, Ranjit and Deccan

iv) Rouging out of infested plants

9) Banded Leaf Sheath Blight

Symptoms: The disease appear on 40-50 days old crop plant. Later it affects flowers. Small brown spots appear on leaves and leaf sheaths. These spots gradually enlarge and brown patches are found in form of bands which is similar to snake's skin (border of the band show brown to dark brown with whitish centre)

Causal organism: *Rhizoctonia solani*

Management

i) Tolerant varieties are Pratap, Kanchan, Maize-3, Shaktiman-1 and Shaktiman-3

ii) Spraying the crop with Validamycin 3%L@ 1250ml/500 litre of water per one hectare

iii) Spray the crop with Hexaconazole 5% EC @ 500ml /500 litre of water per one hectare

Bacterial Diseases

1) Bacterial stalk rot

Symptoms: The symptom first appears on lower nodes of the plant. The infection may confined to one or two inter nodes. The affected tissues become soft but in later stages they turn into dry mass of shredded fibres. The infected plants topple down from 2nd or 3rd basal internode. Two most important characteristics of the disease (i) A sweet fermenting smell is felt around the infected plants. (ii) Toppling down of infected plants from 2nd and 3rd basal internode

Causal organism: *Dickeya zeae*

Syn. *Erwinia chrysanthemi* pv. *zeae*

Management

i) Removal and destruction of diseased plant debris with utmost care.

ii) Growing of resistant or tolerant varieties

iii) Avoidance of water logging conditions in the field.

iv) Stable Bleaching powder @ 12kg/ha should be applied as basal drenching at the time of flowering and second dose should after 10 days of first application.

2) Bacterial leaf stripe

Symptoms: Small oval water soaked spots appear on the leaf surfaces which develop into long stripes. They elongate to cover the entire length of the leaf. Their length usually varies from 1-25cm or more. Later these turn dark brown and papery.

Causal organism: *Acidovorax avenae* sub sp. *avenae*

Syn. *Pseudomonas rubrilineans*

Management

i) Removal and destruction of diseased plant debris.

ii) Growing of resistant or tolerant varieties

iii) Spraying the crop with Streptocycline @ 0.01% after appearance of disease in the field. If necessary second spraying should be given at an interval of 10 days.

Viral diseases

1) Maize Dwarf Mosaic

Symptoms: Infested plants produce long chlorotic streaks along the leaf lamina. Gradually the infected leaves in the whorl become chlorotic. The mosaic patch disappears as the plant matures. The leaves become yellowish green and frequently show blotches.

Causal organism:

Sugarcane mosaic virus

Vector: Aphid

Management

i) Removal and destruction of diseased plant debris with utmost care.

ii) Growing of resistant or tolerant varieties

2) Maize mosaic

Symptoms: Near the base of the young leaf white flecks appear which is associated with whitening of veins. The specks elongate to form discontinuous stripes. These stripes coalesce and form yellow bands on leaf. Severely infected plants may not survive.

Causal organism: Maize mosaic virus

Vector: Leaf hopper

Management

i) Removal and destruction of diseased plant debris with utmost care.

ii) Growing of resistant or tolerant varieties

Minor diseases:

Fungal diseases

Alternaria leaf spot *Alternaria alternata*

Ascochyta leaf and sheath rot *Ascochyta maydis*

Aspergillus ear rot *Aspergillus* sp

Stalk rot *Botryiodiplodia theobromae*

Cercospora leaf spot *Cercospora zea maydis*

Southern rust *Puccinia polyspora*

Yellow leaf spot *Phyllosticta maydis*

Crazy top *Sclerophthora macrospora*

Cob rot *Sphacelia sorghi*

Bacterial diseases

Bacterial stipe *Pseudomonas andropogonis*

Bacterial leaf blight *Pseudomonas avenae*

Chocolate spot *Pseudomonas coronafaciens*

Red stripe *Pseudomonas rubrilineans*

Viral diseases

Maize chlorotic mottle: Maize chlorotic mottle virus (MCMoV)

Maize Dwarf Disease (MSD): Maize dwarf mosaic virus (MDMV)

Maize streak virus disease: Maize streak virus (MSV)

BAJRA (*Pennisetum typhoides*)

Fungal diseases

1) Leaf blast

Symptoms: The disease first appear as light to dark brown boat shaped lesions on the lower surface of leaves. In some cases several spots coalesce to form larger patches. In severe infection leaves may show complete drying.

Causal organism: *Pyricularia setariae*

Management

i) Field sanitation

ii) Use of certified seeds

iii) Growing of resistant varieties

iv) Seed treatment with Carbendazim 50% WP @ 2gm/kg of seed.

v) Spraying of crop with Validamycin (0.2%) or Carbendazim 50% WP (0.15%) two to three times at 7 days interval

2) Rust

Symptoms: Small, orange coloured pustules develop on both the surface of leaves in groups towards distal end of the leaf sheath and stem. In later stage pustules changes to black. In severe infections the plants appear unhealthy and slightly stunted compared with non-infected ones. In severe infection leaves may be completely dried.

Causal organism: *Puccinia penniseti*

Management

i) Field sanitation

ii) Use of certified seeds

iii) Growing of resistant varieties

iii) Spraying of crop with Carboxin 37.5% WP +Thiram 37.5% WP @3gm/litre or Dithane Z-78@ 2g/litre two to three times at 7 days interval after appearance of disease in the field.

3) Green Ear/Downy mildew.

Symptoms: The ears are transformed wholly or partially into loose head of small, twisted green leaf like structure. The infected plants are dwarfed because of shortening of internodes and produce excessive tillers. Infected leaves become yellow or white with yellowish longitudinal streaks extending from the tip to the base. A whitish downy fungal growth can be seen on the lower surface of the leaves and even on the upper surface in severe cases.In many affected plants ear fail to form. Mostly the entire ear is transformed

Causal organism: *Sclerospora graminicola*

Management

i) Field sanitation.

ii) Use of certified seeds.

iii) Treat the seeds with Carbendazim 50%WP@ 2g/kg before sowing.

iv) Rouging of diseased plants should be practised as soon as after the appearance of green ear symptoms in the field.

v) Spraying of crop with Dithane Z-78 @ 0.2%) two to three times at 7 days interval at seedling stage (two to three weeks stage) and also at boot leaf stage.

iv) Growing of resistant varieties like NHB-5,PHB-10 and PHB-14

4) Ergot

Symptoms: The disease occurs at the time of flowering. The honey like pinkish liquid appears on the ears and become brown and sticky. Dark brown coloured elongated sclerotia of the fungus are found.

Causal organism: *Claviceps microcephala*

Management

i) Field sanitation

ii) Follow deep summer ploughing

iii) Use of certified seeds

iv) Two year crop rotation in ergot infested fields

v) The ergot infected ears should be rogued out and destroyed before harvesting of the main crop.

vi) For removal of sclerotia, the seeds should be dipped in 10% solution of common salts then thoroughly washed in clean water and dried in shade. Then seeds are treated with Carboxin 37.5% WP +Thiram 37.5% WP @ 2g/kg before sowing.

vii) Spraying of crop with Mancozeb @ 0.3% at boot leaf stage followed by at least two to three times spraying at 7-10 days interval.

viii) Growing of resistant varieties like NHB-5, PHB-10 and PHB-14

Bacterial Diseases

1) Leaf spot

Symptoms: The water soaked brown spots are formed on the upper surface of the leaves. Several spots coalesce cause blightening of the leaves.

Causal organism: *Xanthomonas penniseti*

Management

i) Field sanitation

ii) Follow deep summer ploughing

iii) Use of certified seeds

2) Leaf blotch

Symptoms: Minute longitudinal streaks develops on the upper surface of leaves. They coalesce to form blotches of orange brown colour. In the late of the season dull brown blotches may also be formed.

Causal organism: *Xanthomonas annamalaiensis*

Management

i) Field sanitation.

ii) Follow deep summer ploughing.

iii) Use of certified seeds.

Viral diseases

1) Mosaic of Bajra

Symptoms: Long chlorotic streaks appear along with veins. Infected plants become stunted. The leaves become yellowish green and frequently show blotches.

Causal organism: Sugarcane mosaic virus

Vector: Aphids (*Myzus persicae*, *Rhopalosiphum maidis*)

Management

i) Removal and destruction of diseased plant debris with utmost care.

ii) Growing of resistant or tolerant varieties

2) Streak of Bajra

Symptoms: Near the base of the young leaves small whitish flecks may appear on one side of the midrib. Chlorotic stripes develop on leaves which run along the entire leaves.These stripes may coalesce form yellow bands. The affected plants produce poorly formed and poorly filled per heads.

Causal organism: Maize streak virus

Vector: Leaf hopper (*Peregrinus maidis)*

Management

i) Removal and destruction of diseased plant debris with utmost care.

ii) Growing of resistant or tolerant varieties

Minor Diseases

Fungal Diseases

Leaf blight: *Drechslera australiensis*

Leaf and ear spot: *Acrothecium penniseti*

Twisted top: *Fusarium moniliforme*

Leaf spot: *Helminthosporium* sp

Leaf spot: *Myrothecium roridum*

Leaf rust: *Puccinia substriata var. penicillariae*

Smut: *Tolyposporium senegalensis*

Smut: *Tilletia ajrekarii*

Viral Diseases

Mosaic: Bajra mosaic virus

BARLEY (*Hordeum vulgare*)

1) Root Rot and Foot Rot

Symptoms: Roots and collar region show browning. Later on, elongated brown spots appear on leaves, glumes and spikelets. Lesions on the internode below the crown or between the crown and seed are typical symptom of infection.

Causal organism: *Ophiobolus sativus*

Management

i) Field sanitation

ii) Follow deep summer ploughing

iii) Use of certified seeds

iv) Seed treatment with Carbendazim 50%WP@ 2g/kg.

v) Growing of resistant varieties

2) Stripe

Symptoms: Affected leaves show yellow stripes on the older leaf blades and sheaths. Later on these linear stripes become brown and the leaf blade become shredded.

Causal organism: *Helminthosporium gramineum*

Management

i) Field sanitation. Rogue out of diseased plants immediately for the infection

ii) Growing of resistant varieties

iii) Use of certified seeds

iv) Soaking the seeds in cold water from 6AM to 10AM on a bright sunny day (Within the period of May-15 to June-15) and exposing them to sun on a cemented floor from 10AM to 5PM.

v) Seed treatment with Carboxin 37.5% WP +Thiram 37.5% WP @3gm/kg of seed

3) Net Blotch

Symptoms: Dark brown blotch develops at or near the leaf blade. The blotches coalesce resulting in dark brown stripes with irregular margins.

Causal organism: *Helminthosporium teres*

Management

i) Field sanitation. Removal of crop residues after harvesting from the field.

ii) Growing of resistant varieties

iii) Use of certified seeds

iv) Seed treatment with Carboxin 37.5% WP +Thiram 37.5% WP @3gm/kg of seed

v) Spraying of crop with Mancozeb 50% WS @ 0.3% two to three times at 7 days interval

4) Spot Blotch

Symptoms: The brown spots appear on the leaves, glumes and spikelets. The spots may coalesce to form blotches. In severe infection the leaves show dry up. In ears grains are affected. Number of grains reduced and they become shrivelled.

Causal organism: *Pyrenophora sativum*

Management

i) Field sanitation. Rogue out of diseased plants immediately for the reducing the infection

ii) Growing of resistant varieties

iii) Use of certified seeds

iv) Seed treatment with Carboxin 37.5% WP +Thiram 37.5% WP @3gm/kg of seed

5) Rust

i) Brown rust or leaf rust

Symptoms: Brown coloured round to oblong pustules appear on leaf blade. In severe infection leaf sheath and culms are affected.

Causal organism:*Puccinia striiformis*

ii) Black rust or stem rust

Symptoms: The reddish brown to dark brown oblong pustules appear on the leaf sheaths and culms and at later stage these pustules become black. In severe infection the grains become shrivelled.

Causal organism: *Puccinia graminis tritici*

iii) Yellow rust

Symptoms: Lemon yellow coloured uredo pustules appear on leaves. These pustules are arranged in stripes and turn black in later stage. In severe infection pustules are formed on the leaf sheaths and glumes.

Causal organism: *Puccinia striiformis*

Management

i) Use of certified seeds

ii) Seed treatment with Carboxin 75%WP @ 2.5 gm/kg of seed

iii) Spraying of crop with Mancozeb 75% WP @0.3% two to three times at 7 days interval after appearance of the disease in the field

iv) Spraying of crop with Tebuconazole 25% WG @1.5ml/litre or Tebuconazole 50% + Trifloxystrobin 25% WG @6g/10litre after appearance of the disease in the field. If possible spraying should be repeated twice or thrice.

6) Loose smut

Symptoms: The panicles are infected. Smutted grains are appeared in place of healthy grains. In early stages they are covered by thin silvery membrane which ruptures easily while the ear emerging out from the sheath. The loose spore masses are easily blown by the wind. Only bare rachis is observed in the field.

Causal organism: *Ustilago nuda*

Management

i) Field sanitation. Rogue out of diseased plants immediately before the rupture of membrane.

ii) Growing of resistant varieties

iii) Use of certified seeds

iv) Soaking the seeds in cold water from 6AM to 10AM on a bright sunny day (within the period of May-15 to June-15)and exposing them to sun on a cemented floor from 10AM to 5PM.

v) Seed treatment with Carboxin 75% WP @ 2.5gm/kg or Carboxin 37.5% WP +Thiram 37.5% WP @3gm/kg of seed

7) Covered smut

Symptoms: The affected ears may emerge at the same time as healthy ears. The disease is recoginzed after the emergence of ears. Every ear in a diseased stool and every grain in a diseased ear is affected. The black spore mass of this smutted grains remains covered by adhering membramne of the grain and basal part of the glumes. The awns remain intact though they may sometimes found partly damaged. The spore masses held together by fatty substances. The smutted ears become prominent when the crop reaches maturity. The spores are released at the time of threshing.

Causal organism: *Ustilago hordei*

Management

i) Use of certified seeds

ii) Growing of resistant varieties. In north Indian conditions the resistant varieties are Kalyansona,WG-307, C-302 and PV-18.

iii) Rogue out of diseased plants immediately before the rupture of membrane.

iv) Soaking the seeds in cold water from 6AM to 10AM on a bright sunny day (within the period of May-15 to June-15) and exposing them to sun on a cemented floor from 10 AM to 5 PM.

v) Seed treatment with Carboxin 37.5% WP +Thiram 37.5% WP @3gm/kg of seed.

8) Powdery mildew

Symptoms: Powdery patches appear on the leaves and other green parts. The superficial mycelium is grey in colour. Later on these patches become dark coloured are covered with black dot like cleistothecia. The leaves become yellow, brown and gradually drying out

Causal organism: *Erysiphe graminis*

Management

i) Field sanitation

ii) Follow deep summer ploughing

iii) Use of certified seeds

iv) Spray the crop with Sulphur 80% WDG/WG @ 5g/litre

Viral diseases

1) Barley Stripe Mosaic

Symptoms: The affected plants remain stunted show mosaic mottled leaves which are sometimes associated with flecking. In later stage the colour of leaves become yellow.

Causal organism: Barley stripe mosaic virus (Hordei viruses) remains viable for 8 years.

Management

i) Field sanitation. Rogue out of diseased plants from the field.

ii) Growing of resistant varieties

iii) Use of certified seeds

2) Barley Yellow Dwarf

Symptoms: Affected plants remain dwarfed. At first faint yellow green blotches occur near the leaf tip later the colour of leaves become bright yellow.

Causal organism: Barley yellow dwarf virus (Luteoviruses)

Vectors: Aphids (*Rhopalosiphum maidis, Macrosiphum granarium*)

Management

i) Field sanitation. Rogue out of diseased plants from the field.

ii) Growing of resistant varieties

iii) Use of certified seeds

Nematode diseases

1) Molya

Symptoms: Infected plants show stunted growth and general chlorosis. Tillering is reduced. Infected plants flower prematurely. The growing point of root is inhibited. New roots are generated showing characteristic rosetting effect. The affected plants become susceptile to drought conditionsas there are no roots penetrating into the deeper soil layers. After 4-6 weeks of sowing, i.e.by the end of November or early December slight swelling near the root tips can be detected. White glistening female nematodes can be seen adhering to roots. These cysts turn brown and may remain attached to roots or fall off in the soil after the roots decay. In newly infected fields the disease appears in patches and these patches gradually increase with continuous cropping of either balrey and wheat.

Causal organism: *Heterodera avenae*

Management

i) Crop rotation with non-host crops like gram, carrot, onion, mustard and fenugreek.

ii) Deep Summer ploughings for two to three times also reduce the nematode infestation.

iii) Green manuring also reduce the nematode population.

iv) Soil application of Carbofuran@ 2kg ai/ha before planting reduce the nematode population and increase the yield.

Minor diseases

Fungal diseases

Leaf blotch *Alternaria alternata*

Leaf blight *Helminthosporium catenarium*

Water-soaked leaf spot *Drechslera hawaiiense*

Root rot and foot rot *Fusarium culmorum*

Ascochyta leaf spot *Ascochyta graminea*

Take all *Gaeumannomyces graminis* var *tritici*

Cephalosporium stripe *Cephalosporium gramineum*

Ergot *Claviceps purpurea*

Anthracnose *Colletotrichum graminicola*

Net blotch *Heleminthosporium teres*

Blast *Pyricularia grisea*

Root rot *Sclerotium rolfsii*

Wilt, stem and root rot: *Ozonium taxanum* var. *parasiticum*

White rot *Pellicularia rolfsii*

Browning root rot *Pythium graminicolum*

Covered smut *Ustilago hordei*

Loose smut *Ustilago nuda*

Tundu *Corynebacterium tritici*

Bacterial diseases

Basal glume rot *Pseudomonas atrofaciens*

Bacterial stripe *Pseudomonas striafaciens*

Viral diseases

Mosaic

American wheat striate mosaic virus (WSMV)

Easternwheat striate mosaic virus (EWSMV)

WHEAT (*Triticum aestivum*)

Fungal diseases

1) Pythium Root rot

Symptoms: The seedlings may show damping off under low soil temperature. Soft rot of roots and root lets occur. Seedlings become stunted and look pale green. followed by wilting. In older plants the collar region become discoloured, soft and wilting may occur.

Causal organism: *Pythium graminicolum*

Management

i) Field sanitation should be strictly followed.

ii) Crop rotation.

iii) Balanced fertilizer application.

2) Sclerotial Disease

Symptoms: Near the base of the plant white mycelial growth is observed on which sclerotia develop.

Causal organism: *Pellicularia rolfsii*

Management

i) Field sanitation should be strictly followed.

ii) Crop rotation

iii) Balanced fertilizer application.

3) Leaf or brown rust

Symptoms: Small round light yellowish brown pustules appear on the leaves being rare on leaf sheaths and stalks which later turn black covered by the epidermis. The uredopustules are bright orange in colour, bigger in size than the uredia of yellow rust and are present in small clusters or irregularly scattered. This rust occurs widely cause serious yield loss. In severe infection the plant matures early having poor root system and produce shrivelled and light grains.

Causal organism: *Puccinia recondita*

4) Yellow or Stripe rust

Symptoms: The disease usually appears earlier the stem rust before grains are formed. The uredopustules are oval in shape,lemon yellow in colour and are arranged in rows on the leaves. In severe infection they are formed on the leaf sheaths stalks and glumes. In severe infection these rows are not distinct as the large patches of the leaf become covered. In severely attacked fields the ground may appear orange or red due to falling of enoromous quantity of rust spores on it.

The telia appear as black patches or spots on the under surface of the leaf. Like uredia the telia are often arranged in rows. They may appear on other parts of the plant. They do not break hrough the epidermis and remain as a flat black crust.

Causal organism: *Puccinia striiformis*

5) Black Rust or Stem Rust

Symptoms: The elongated reddish brown uredial may be about 6 mm or more in length pustules are found mainly on stem and leaf sheaths. The infection is less in leaves and earheads. When the crop matures dark black telial pustules are developed.

The black rust appears in endemic form in South India, Central India and East India where high temperature prevails during wheat growing season.

The first symptom of rust infection is flecking on leaves, leaf sheaths, culms and floral structures. These flecks develop as oblong, reddish brown pustules, frequently merging into one another and finally rust expose a mass of brown uredospores. In case of severe infection the diseased plants are stunted produce small spikes and grains or no grain at all.

Causal organism: *Puccinia graminis tritici*

Management

i) Growing of resistant varieties

ii) Use of certified seeds

6) Powdery mildew

Symptoms: Whitish to greyish powdery patches appear on the upper surface and then on the under surface of leaves. Such symptoms also appear on leaf sheaths and floral parts. At the later stage the powdery growth become dark grey and finally black causing characteristic lesions of varying sizes on affected parts on which numerous small round dark cleistothecia develop.

Causal organism: *Erysiphe graminis* f. sp. *tritici*

Management

i) Field sanitation.

ii) Growing of moderately resistant varieties, i.e. NP 710, NP718, K-53, E-750 and C-591

iii) Use of certified seeds.

iv) Use of 1% Sodium thiosulphate as a foliar spray

v) Spraying of Sulphur 80% WG @5.0 gm/litre after appearance of the disease in the field.

If necessary the spraying may be repeated twice or thrice at an interval of 7 days.

7) Alternaria leaf blight

Symptoms: The disease first appears in the field when the crop is 7-8 weeks old. Initially the older leaves are infected showing reddish brown localized oval spots which enlarge to become as irregular with bright yellow margin. Several spots may coalesce and cause blightening of leaves. Heavily infected fields give burnt appearance even from a distance. The plants are killed due to severe defoliation and stem infection.

Causal organism: *Alternaria triticina*

Management

i) Seed treatment with Carbendazim 50%WP 2-2. 5g/kg or Carboxin 17.5% WP +Thiram 17.5% WP @ 3gm/kg of seed.

ii) Spraying of crop with Mancozeb (0.2%) twice/thrice at 7 days interval after appearance of the disease in field

8) Drechslera or Helminthosporium leaf spot

Symptoms: The fungus attacks the plant at all stages of growth starting from seedling to matured stage. Oval to oblong spots of 1-2 mm size appear on leaf blade. In severe cases the sheath is affected and spots, of increase in size and become dark brown and cause blightening of seedlings. In older plants dry straw patches develop with brown margin surrounded by yellow halo where the central portion is irregular.

Causal organism: *Helminthosporium sativum*

Management

i) Field sanitation should be strictly followed. Infected crop stubbles should be collected and deep buried in soil.

ii) Seed treatment with Carbendazim 50% WP@ 2.5g/kg or Carboxin 17.5% WP +Thiram 17.5% WP @3gm/kg of seed.

iii) Spraying of crop with Mancozeb (0.2%) twice/thrice at 7 days interval after appearance of the disease in field

9) Loose smut

Symptoms: In diseased plant the ear heads are emerged earlier. Almost every grain in the infected ear is converted into black powder (which are actually the spores of the fungus). Prior to emergence each smutted spikelets are covered by a silvery membrane which breaks while the ear emerges. The powdery masses of spores is blown off by wind leaving behind the only central rachis

Causal organism: *Ustilago tritici*

Management

i) Rogue out of diseased plants immediately before the rupture of membrane.

ii) Growing of resistant varieties. In north Indian conditions the resistant varieties are Kalyansona,WG3 07,C 302 and P 18.

iii) Use of certified seeds

iv) Soaking the seeds in cold water from 6AM to 10AM on a bright sunny day (Within the period of May-15 to June-15) and exposing them to sun on a cemented floor from 10 AM to 5 PM.

v) Seed treatment with Carboxin 37.5% WP +Thiram 37.5% WP @3gm/kg of seed

10) Karnal Bunt

Symptoms: In this disease some grains in the panicle are partially or wholly converted into black powdery masses. When the infected grains mature the outer glumes spread out and the inner glumes expand, expose the bunted grains. The embryo tissue is generally not destroyed. The bunt balls are at first enclosed by the pericarp, but when it bursts the masses of spores are exposed resulting a characteristic fishy odour in the field. The fungus is both seed borne and soil borne and at the time of flowering the infection occurs by air borne sporidia.

Causal organism: *Neovossia indica*

Management

i) Growing of resistant varieties like Kalyansona, PV-18, HD 2012, HD 4513, HD 4519, HI7715, HI 7719

ii) Use of certified seeds

iii) Seed treatment with Carboxin 17.5% WP +Thiram 17.5% WP @3gm/kg of seed

iv) Avoid excess irrigation at the time of flowering.

11) Hill bunt

Symptoms: The disease can be identified after the emergence of ears. The smutted ears are darker green colour and usually remain green for longer period than normal ear. The grains are filled up with a soft black coloured paste like substance. In later stage they become black powder. All the grains in the ear are affected and transformed into bunt balls. The spore balls are crushed at time of harvest and the black spores are released. Ultimately the heathy grains are contaminated by these black bunt spore.

Causal organism: *Tilletia caries* and *T. foetida*

Management

i) Growing of resistant varieties like Kalyansona, PV18,HD 2012,HD 4513, HD 4519, HI 7715, HI 7719.

ii) Use of certified seeds

iii) Seed treatment with Carboxin 37.5% WP +Thiram 37.5% WP @3gm/kg of seed

12) Leaf Blotch

Symptoms: Light green to yellow spots appear between the veins of the lower leaves. These spots enlarge to form reddish brown lesions that are often partly surrounded by a yellow margin. At the late stage these spots turn light brown to almost white with small dark black dots. Heavily infected young leaves dry up prematurely. In extreme cases the plants become stunted and ears may not develop or if develop they may sterile.

Causal organism: *Septoria tritici*

Management

i) Field sanitation should be strictly followed. Infected crop stubbles should be collected and deeply buried in soil.

ii) Crop rotation should be followed for at least three years.

iii) Growing of resistant varieties.

iv) Seeds from disease free fields should be used.

v) Seed treatment with Carbendazim 50% WP @0.15%.

vi) Spraying of crop with Mancozeb 50 % WS @ 0.3%.

13) Glume Blotch

Symptoms: Small, linear or long blotches appear on the nodal tissue and floral bracts. The blotches are dark brown in colour on which black dot like structure, i.e. pycnidia are present.

Causal organism: *Septoria nodorum*

Management

i) Field sanitation should be strictly followed.

ii) Crop rotation should be followed for at least three years.

iii) Growing of resistant varieties.

iv) Seeds from disease free fields should be used.

v) Seed treatment with Carbendazim 50% WP @0.15%.

vi) Spraying of crop with Mancozeb 50% WS @0.3%.

14) Flag smut

Symptoms: The symptoms are usually found on the leaves and leaf sheaths. In smut the lesions appear as long light green stripes which soon become light green streaks. Rolling and twisting of leaves and the leaves split longitudinally. The infection is more towards the upper portion of the leaf blade than at the base. Infected plants become shorter and produce tillers profusely. If the infection occurs at seedling stage, the plant shows severely stunting,the leaves become twisted and no ear formation. In highly susceptible varieties dwarfing of plants and arresting of development of inflorescences occur.

Causal organism: *Urocystis agropyri*

Management

i) Crop rotation should be followed for at least three years.

ii) Hot weather cultivation should be practised.

iii) Early sowing of the crop, infected crop stubbles should be collected and burnt or deeply buried in soil.

iv) Growing of resistant varieties, i.e. Pusa 4.

v) Use of certified seeds

vi) Seed treatment with Carboxin 37.5% WP +Thiram 37.5% WP @3gm/kg of seed.

Bacterial and Nematode disease complex

1) Tundu or Yellow Ear Rot

Symptoms: The disease appears at the time of maturity of the plants. The stem and inflorescence of the plants are affected. Yellow slime appear on stem and inflorescence which dries up to form sticky yellow layers. Curling and twisting of the spikes occur. Most of the grains are replaced by nematode galls and these galls are the carrier of the causal organism, i.e. bacterium. If the nematode infection occurs at young seedling stage twisting and distortion of the leaves take place. In adult plants twisting of ear heads and formation of galls in place of normal grain occur.

Causal organism: *Corynebacterium tritici and Anguina tritici*

Management

i) Field sanitation should be strictly followed. Infected crop debris should be collected and deep buried in soil.

ii) Crop rotation should be followed for at least three years.

iii) Growing of resistant varieties.

iv) Seeds from disease free fields/ gall free seeds or certified seeds should be used.

v) In disease prone areas the seeds should be treated with 20% salt solution and stirred vigorously. Hence the affected seeds along with galls may float on the surface which can be separated with the help of sieves. The healthy seeds which settle down should be collected and washed thoroughly followed by drying. Then can be sown in the field.

Virus diseases

1) Wheat Yellow Dwarf

Symptoms: Affected plants remain dwarfed. At first faint yellow green blotches occur near the leaf tip later the colour of leaves become bright yellow.

Causal organism: Barley yellow dwarf virus (Luteoviruses)

Management

i) Field sanitation. Rogue out of diseased plants from the field.

ii) Growing of resistant varieties

iii) Use of certified seeds

Nematode diseases

1) Ear Cockle

Symptoms: The infested plants remain stunted and the leaves show generally twisted, crinkled and rolled. The ears are shorter and thicker than healthy ones. The grains are replaced by dark coloured hard galls which are full of nematodes.

Causal organism: *Anguina tritici*

Management

i) Field sanitation should be strictly followed. Infected crop debris should be collected and deeply buried in soil.

ii) Crop rotation should be followed for at least three years.

iii) Growing of resistant varieties.

iv) Certified seeds or seeds from disease free fields should be used.

v) The seeds can be made gall free by treating with 20% brine solution.

2. Molya

Symptoms: The infected plants show stunted growth. The length of the main root reduces become bunchy bearing small galls. In case of severe infestation the seedling may fail to come out and if they come out they remain stunted. In case of early infction the plants produce short stalks and panicles resulting poor yield.

Causal organism: *Heterodera avenae*

Management

i) Growing of resistant varieties

ii) Use of certified seeds

iii) Crop rotation with non-host crops like gram, carrot, onion, mustard and fenugreek for two years.

iv) Deep Summer ploughings for two to three times also reduce the nematode infestation.

v) Green manuring also reduce the nematode population.

vi) Soil application of Carbofuran @ 2kg ai/ha before planting reduce the nematode population and increase the yield.

Minor diseases

Fungal diseases

Anthracnose *Colletotrichum graminicola*

Cephalosporium stripe *Cephalosporium gramineum*

Downy mildew *Sclerophthora macrospora*

Ergot *Claviceps purpurea*

Leaf spot *Ascochyta sorghi*

Ear blight *Fusarium avenaceum*

Seedling blight *Fusarium culmorum*

Root rot *Fusarium equiseti*

Root rot *Fusarium graminearum*

Take all *Gaeumannomyces graminis var. tritici*

Sclerotial disease *Pellicularia rolfsii*

Bacterial diseases

Bacterial leaf blight *Pseudomonas syriange*

Basal gum rot *Pseudomonas atrofaciens*

Black chaff *Xanthomonas transluscens* pv. *undulosa*

Viral diseases

Mosaic Barley stripe mosaic virus (BSMV)

Wheat streak mosaic virus (WSMV)

SORGHUM (*Sorghum bicolor*)

Fungal diseases

1) Seed rot, seedling blight and leaf blight

Symptoms: The fungus cause death of the plants in seedling stage and blightening of leaves in growing stage. Due to seed infection there may be poor stand in the field. In some cases the collar region of seedlings are attacked which ultimately cause wilting followed by death. Small reddish purple or tan spots appear on leaves which may enlarge coalesce and blighted the leaves. In growing stage older leaves are affected. Greyish or tan elliptical lesions of several inches long appear on the leaves which turn into straw or brown coloured with dark margins in later stage.

Causal organism: *Helminthosporium turcicum*

Management

ii) Growing of resistant varieties.

iii) Use of certified seeds

iv) Seed treatment with Carboxin 37.5% WP + Thiram 37.5% WP @ 3gm/ kg of seed

2) Anthracnose

Symptoms: The disease attack the plants at both seedling and adult stage. Small reddish purple and brownish spots appear on the lower leaves. The centre of the spots become whitish. In later stages small black dot like fruiting bodies appear. Plants show defoliation. The severe infection leads to death of the plants

Causal organism: *Colletotrichum graminicola*

Management

i) Growing of resistant varieties.

ii) Use of certified seeds

iii) Seed treatment with Carboxin 37.5% WP + Thiram 37.5% WP @3gm/ kg of seed

iv) Spraying of crop with Mancozeb @ 0.3%

3) Stalk rot

Symptoms: There is emergence of twisted and rolled leaves in infected plants. The upper inter nodes may be shorter and the dark lesions develop on outer side of the node. The pith of infected stalk is water soaked and later on the infected stalk tissue turns deep red in colour. Flowering is rare in infected plants

Causal organism: *Fusarium moniliforme*

Management

i) Removal of stubbles as the fungus survive on crop debris

ii) Clean cultivation

iii) Growing of resistant varieties

4) Charcoal rot

Symptoms: Both seedlings and adult plants are infected. Small and black sclerotia are seen on infected tissues. Affected plants show lodging in severe infection.

Causal organism: *Macrophomina phaseoli*

Management

i) Removal of stubbles as the fungus survive on crop debris

ii) Clean cultivation

iii) Growing of resistant varieties

iv) Crop rotation with non-host crops

5) Cercospra leaf spot

Symptoms: Dark purple or reddish spots appear on both the surface of leaves. The centre of spot becomes tan. Severe infection may cause drying of the leaves.

Causal organism: *Cercospora sorghi*

Management

i) Growing of resistant varieties.

ii) Use of certified seeds

iii) Seed treatment with Carboxin 37.5% WP +Thiram 37.5% WP @3gm/kg of seed.

iv) Spraying of crop with Mancozeb @0.3%

6) Zonate leaf spot

Symptoms: Small red to brown water soaked lesions appear on leaves which turn into dark coloured irregular lesions. Sclerotia bodies appear on them. Severe infection cause defoliation and death.

Causal organism: *Gloeocercospora sorghi*

Management

i) Growing of resistant varieties.

ii) Use of certified seeds

iii) Seed treatment with Carboxin 37.5% WP +Thiram 37.5% WP @3gm/kg of seed

iv) Spraying of crop with Mancozeb @0.3%

7) Ascochyta leaf spot/Rough leaf spot

Symptoms: Small, circular to oblong light coloured spots with well defined margin develop on the leaves. These spots enlarge and coalesce to form large blotches. Black raised pycnidia develop on these blotches. These lesions become yellow brown or purple red colour. The matured leaves become necrotic.

Causal organism: *Ascochyta sorghina*

Management

i) Growing of resistant varieties.

ii) Use of certified seeds

iii) Follow the crop of rotation.

8) Sooty Stripe

Symptoms: Water-soaked oblong reddish spots appear on old leaves. The leaves turn bright yellow. Superficial black sclerotia formed in centre of lesions give sooty appearance.

Causal organism: *Ramulispora sorghi*

Management

i) Seed treatment Carboxin 37.5% WP +Thiram 37.5% WP @3gm/kg of seed.

ii) Crop rotation

9) Ergot:

Symptoms: The grains are replaced by fungal growth. The typical honey dew secretion takes place from the affected spikelets

Causal organism: *Sphacelia sorghi*

Management

i) Growing of resistant varieties.

ii) Use of certified seeds

iii) Seed treatment with Carboxin 37.5% WP +Thiram 37.5% WP @3gm/kg of seed.

iv) Spraying of crop with Mancozeb 50% WS @0.3%

10) Grain smut

Symptoms. The disease appears at the time of grain formation. The smut sori are formed in place of healthy grains. These are generally larger than the normal grains.

Causal organism: *Spacelotheca sorghi*

Management

i) Growing of resistant varieties.

ii) Use of certified seeds

iii) Seed treatment with Carboxin 37.5% WP +Thiram 37.5% WP @3gm/kg of seed

iv) Spraying of crop with Mancozeb 50% WS @0.3%

11) Loose smut

Symptoms: The infected plants are thinner than the normal ones. They produce more tillers and usually flower earlier than normal healthy plants. The affected ears look leafy or leathery structures. In the infected ears all spikelets become smutted and malformed.

Causal organism: *Sphacelotheca cruenta*

Management

i) Crop rotation

ii) Field sanitation.

iii) Rogue out diseased plants immediately before the rupture of membrane.

iv) Growing of resistant varieties

v) Use of certified seeds

12) Head smut

Symptoms: At the time of flowering the disease appears. In infected plants the inflorescence is totally damaged. In some cases the partial damage has been occurred. The entire inflorescence is converted into a big sorus of about 1cm length and 5cm breadth. In early stages it is covered by a thin white membrane but during emergence through the boot leaf the membrane ruptures and large amount of black coloured spore masses are exposed which are wind blown.

Causal organism: *Spacelotheca reliana*

Management

i) Crop rotation

ii) Field sanitation.

iii) Rogue out diseased plants immediately before the rupture of membrane.

iv) Growing of resistant varieties

v) Use of certified seeds

13) Long smut

Symptoms. A few grains in the ear are converted into sorus which is covered by a whitish or dull yellow, fairly thick membrane. Each sorus is long cylindrical and slightly curved and rupture by mechanical means to release the brownish green spore balls.

Causal organism: *Tolyposporium ehrenbergii*

Management

i) Crop rotation

ii) Field sanitation.

iii) Rogue out diseased plants immediately before the rupture of membrane.

iv) Growing of resistant varieties

v) Use of certified seeds

Bacterial diseases

1) Red leaf blotch

Symptoms: The water soaked lesions develop on the younger leaves which are surrounded by chlorotic halo. The affected plants are become stunted. Ears do not emerge if emerge the grain size become small and chaffy.

Causal organism: *Xanthomonas rubrisorghi*

Management

i) Removal of stubbles as the pathogens survive on crop debris.

ii) Clean cultivation.

iii) Growing of resistant varieties.

iv) Crop rotation with non-host crops.

2) Bacterial Leaf Stripe

Symptoms: Reddish brown narrow lesions develop on the leaves. Gradually the lesions turn dark red and extend along the veins. These lesions coalesce and form linear irregular stripes covering large areas of the sleaf, bacterial exudates may appear on the under surface of the leaves. Gradually these exudates dry and form red crusts or scales.

Causal organism: *Pseudomonas sorghicola*

Management

i) Growing of resistant varieties

ii) Crop rotation with non-host crops

Viral diseases

1) Mosaic

Symptoms

Irregular mottling of light and dark green areas with longitudinal yellow and white streaks are observed on upper two to three leaves. Severe infection may cause death of the plants.

Causal organism: Maize dwarf mosaic virus (MDMV)

Management

i) Clean cultivation

ii) Growing of resistant varieties

iii) Spraying of crop with insecticides after appearance of the disease in the field

2) Chlorosis

Symptoms: Longitudinal chlorotic bands appear on the blades of young leaves. Later the leaves become complete chlorotic. The leaves are bunched together at the apex. Early flowering is noticed in infected plants. Plants remain dwarf, chlorotic and sterile.

Causal organism: Sugarcane mosaic virus (SCMV)

Management

i) Clean cultivation

ii) Growing of resistant varieties

iii) Spraying of crop with insecticides after appearance of the disease in the field

Phanerogamous parasites

Striga

Symptoms: The plants remain stunted due to infestation by striga. These are partial root parasite and appears above ground level around the plants when they were fully gown. These parasites produce more than thousands minute seeds per plant within 25-30 days after emergence.

Causal organism

i) *Striga asiatica*

ii) *S. densiflora*

Management

i) Growing of resistant varieties.

ii) The shoot of the parasites should be pulled out before seed setting

iii) Spraying of crop with weedicide, i.e. 2,4-D

Minor diseases

Fungal diseases

Leaf spot: *Helminthosporium sativum*

Midrib spot: *Colletotrichum falcatum*

Target leaf spot: *Helminthosporium sorghicola*

Southern leaf blight *H.maydis*

Rhizoctonia blight and rot: *Rhizoctonia solani*

Seed and seedling blight: *Pythium* sp.

Green ear: *Sclerospora graminicola*

Curvularia kernel rot: *Curvularia lunata*

Sclerotial rot: *Sclerotium rolfsii*

Phoma leaf spot: *Phoma insidiosa*

Leaf rust *Puccinia sorghi*

Tar leaf spot *Phyllochora sorghi*

Bacterial diseases

Yellow leaf blotch *Pseudomonas* sp

Bacterial spot *P. syringae*

Bacterial leaf stripe *P. sorghicola*

OAT (*Avena sativa*)

Fungal diseases

1) Seedling blight

Symptoms: Poor germination and also plant stand reduce. Seedlings turn yellow and collapse in severe infection.

Causal organism

i) *Rhizoctonia* sp

ii) *Fusarium* sp

iii) *Helminthosporium* sp

Management

i) Seed treatment with Carboxin 37.5% WP +Thiram 37.5% WP @3gm/kg of seed

2) Leaf blotch

Symptoms: Oblong to elongate light reddish-brown spots appear on newly emerged leaves of the seedlings. On older leaves elongated greyish brown lesions are found. These lesions coalesce to form large dark brown to grey blotches. In severe infection the tissues may become necrotic dry up.

Causal organism: *Pyrenophora avenae*

Management

i) Seed treatment.

ii) Crop rotation with non-cereal crops.

3) Stem Rust

Symptoms: Reddish brown to dark brown pustules develop on the stems, leaf sheaths, blades and panicles of the plant. At the maturing stage of the plant pustule become black on stems and sheaths.

Causal organism: *Puccinia graminis* f. sp. *avenae*

Management

i) Seed treatment with Carbendazim 50%WP @ 2gm/kg cr Carboxin 37.5%WP + Thiram 37.5% WP @ 3 gm/kg of seed

ii) Use of certified seeds

iii) Spraying of crop with Mancozeb 50% WS @ 0.3% two to three times at 7 days interval after appearance of the disease in the field

iv) *Vulpia myuros* and *Birza minor* are the common wild hosts. They are the alternate host which are found in Nilgiri hills. Alternate hosts should be eradicated.

4) Crown Rust

Symptoms: Bright orange yellow round to oblong uredopustules develop on the leaves. Similar pustules are also formed on sheaths,stems and panicles. Later on, teleutospores are formed which remain covered indefinitely by the epidermis revealing a raised appearance.

Causal organism: *Puccinia coronata* var. *avenae*

Management

i) Seed treatment with Carboxin 37.5% WP +Thiram 37.5% WP @3gm/kg of seed

ii) Use of certified seeds

iii) Spraying of crop with Mancozeb 50% WS@ 0.3% two to three times at 7 days interval after appearance of the disease in the field

5) Loose smut

Symptoms: The grains in the ears are transformed into a loose mass of black powdery spores which are enclosed in a thin whitish grey membrane which ruptures soon after the panicles emerge. Smutted plants are shorter than healthy ones.

Causal organism: *Ustilago avenae*

Management

i) Rogue out of diseased plants immediately before the rupture of membrane.

ii) Growing of resistant varieties.

iii) Use of certified seeds

iv) Soaking the seeds in cold water from 6AM to 10AM on a bright sunny day (within the period of May-15 to June-15) and exposing them to sun on a cemented floor from 10AM to 5PM.

v) Seed treatment with Carboxin 37.5% WP +Thiram 37.5% WP @3gm/kg of seed

6) Covered smut

Symptoms: The grains in the ears are replaced by the black mass of smut spores which are remain enclosed by a whitish grey membrane. The membrane remain intact and sori are broken during ripening.

Causal organism: *Ustilago kolleri*

Management

i) Roguing out diseased plants immediately before the rupture of membrane.

ii) Growing of resistant varieties.

iii) Use of certified seeds.

v) Seed treatment with Carboxin 37.5% WP +Thiram 37.5% WP @3gm/kg of seed.

Minor diseases

Fungal diseases

Seedling rot *Pythium debaryanum*

Anthracnose *Colletotrichum graminicola*

Seed and seedling blight *Fusarium* sp

Seed and seedling blight *Rhizoctonia* sp

Victoria blight *Helminthosporium victoriae*

Foot rot *Fusarium culmorum*

Root rot *Fusarium roseum*

Root rot *Sclerotium rolfsii*

Root rot and stem rot *Sclerotinia sclerotiorum*

Snow mould *Fusarium nivale*

Cephalosporium stripe *Cephalosporium gramineum*

Ergot *Claviceps purpurea*

Kernel rot *Curvularia lunata*

Leaf and glume blight *Drechslera sorghicola*

Leaf blight *Drechslera tetramera*

Leaf spot *Helminthosporium rostratum*

Leaf spot *Helminthosporium sativum*

Leaf spot *Phyllosticta sorghina*

Leaf spot *Phyllosticta sorghiphilla*

Leaf rust *Puccinia sorghi*

Bacterial diseases

Halo blight *Pseudomonas coronafaciens*

Bacterial stripe blight *Pseudomonas striafaciens*

Viral diseases

Mosaic: Wheat streak mosaic virus (WSMV)

American wheat striate mosaic virus (AWSMV)

African cereal streak virus (MSV)

2

Diseases of Millets

FINGER MILLET (*Eleusine coracana*)

Fungal diseases

1) Blast

Symptoms: The above ground parts (leaf, leaf sheath, stems) of the plant exihibit the symptoms. The seedlings and adult plants are affected by the pathogen. Eye-shaped spots appear on leaf and leaf sheath having dark brown margin with ashy centre. These spots coalesce to form large patches and the leaves are blighted. Affected leaves wither prematurely and the seedlings may be killed. Lesions appear on the leaf blades, leaf sheaths and culms on adult plants and cause blightening.

Neck infection: The base portion of the fingers are affected showing brown to dark brown discolouration and ultimately lead to breaking and hanging down of the fingers.

Causal organism: *Pyricularia grisea*

Management

i) Seed treatment

ii) Spraying the crop with Tricyclazole (0.1%).

iii) In endemic areas one spraying should be given at the nursery or seedlings stage, two sprayings in growing stage and one or two sprayings at finger emergence stage are recommended.

2) Charcoal rot

Symptoms: Roots and lower portion of the stems are affected. Larger lesions are formed on the stem where small black sclerotia are formed at the later stage of growth. Affected plants show rotting of the stem.

Causal organism: *Macrophomina phaseolina*

Management

i) Clean cultivation

ii) Use of certified seeds

iii) Spraying of Copper oxychloride @ 3g/litre twice or thrice at 7 days interval after appearance of the disease in the field.

3) Helminthosporium blight

Symptoms: Both the seedlings and adult plants are affected by the fungus. Minute, oval light brown lesions appear on the young leaves. These lesions enlarge and coalesce to form large patches of infection. Lesions appear on leaf, leaf sheath and culms of the adult plants. The infection in the neck region is revealed brown to dark brown discolouration followed by breaking and hanging down of ears.

Causal organism: (i). *Cochliobolus nodulosum,* (ii). *Helminthosporium tetramera* and (iii) *H. leucostylum*

H. nodulosum causes seedling blights and root rots where as *H. leucostylum* causes seedling blight stripes or shredding of leaves and head blight.

Management

i) Field sanitation . Infected crop stubbles should be collected and deeply buried in soil.

ii) Seed treatment with Carbendazim 50%WP @ 2-2.5g/kg or Carboxin 37.5% WP +Thiram 37.5% WP @ 3 gm/kg of seed.

iii) Spraying of crop with Mancozeb 50%WS(0.2%) twice/thrice at 7 days interval after appearance of the disease in field

Bacterial diseases

1) Bacterial leaf spot

Symptoms: The typical symptoms appear on both the surfaces of leaves as light yellowish to brown spots. These are linear spreading along the veins. As the disease advances leaf blade gets shredded by splitting along the streaks.

Causal organism: *Xanthomonas elusineae*

Management

i) Spray Copper oxychloride (0.2%).

2) Bacterial Blight

Symptoms: Water-soaked linear streaks appear on leaves which run parallel to the midrib, then these translucent streaks develop into lesions and ultimately the entire leaf may turn brown and wither.

Causal organism: *Xanthomonas coracana*

Management

i) Spray the Copper oxychloride (0.2%) when the disease appears in the field.

ii) Repeat it twice or thrice if necessary.

Viral diseases

1) Mosaic disease

Symptoms: The symptom is characterized by the chlorosis followed by mosaic, mottling and severe stunting. Affected plants rarely produce ears where the panicles are mostly chaffey.The diseased plants wither prematurely.

Causal organism: Virus

Management

i) Spraying the crop with insecticides to control the insect vector

Minor diseases

Fungal diseases

Charcoal rot *Macrophomina phaseoli*

Curvularia leaf spot *Curvularia lunata*

Damping off *Pythium aphanidermatum*

Leaf spot *Drechslera nodulosa*

Leaf spot *D.rostrata*

Cercospora leaf spot *Cercospora eleusinis*

Root,stem and foot rot *Sclrotium rolfsii*

Root,stem and foot rot *Pellicularia rolfsii*

Ovary Smut *Ustilago eleusinis*

Bacterial diseases

Yellow leaf blotch *Pseudomonas* sp.

BARANYARD MILLET or SAWAN (*Echinochloa frumentacea*)

Fungal diseases

1. Leaf Blight

Symptoms: Dot like numerous minute spots appear on both sides of the leaves and also on leaf sheaths. Later on these spots coalesce cause blightening of leaves.

Causal organism: *Helminthosporium frumentacei*

Management

i) Field sanitation should be strictly followed. Infected crop stubbles should be collected and deep buried in soil.

ii) Seed treatment with Carbendazim 50%WP @ 2-2. 5g/kg or Carboxin 37.5% WP +Thiram 37.5% WP 3gm/kg of seed.

iii) Spraying of crop with Mancozeb 50%WS(0.2%) twice/thrice at 7 days interval after appearance of the disease in field

2. Downy mildew

Symptoms: Light yellow bands appear on leaves which turn into white in later stage. Finally the leaves dry, earheads become chaffy.

Causal organism: *Sclerophthora macrospora*

Management

i) Field sanitation.

ii) Growing of resistant varieties

iii) Use of certified seeds

iv) Seed treatment with Carboxin 37.5% WP +Thiram 37.5% WP @3gm/kg of seed

3. Smut

Symptoms: In the ears grains are affected. Infected grains contain black powder smut spores. In case of *Ustilago frumentacei* infection the grain size is larger than normal grains and in case of *U paradoxa* the size of the grain is not affected.

Causal organism: *Ustilago panici frumentacei*

U. paradoxa

Management

i) Rogue out of smutted ears and destroyed them

ii) Growing of resistant varieties

iii) Use of certified seeds

iv) Seed treatment with Carboxin 37.5% WP +Thiram 37.5% WP @3gm/kg of seed

4. Shoot smut

Symptoms: Swellings are observed in infected stems and also in axil of leaves. The infected inflorescence is converted to swelling/ gall like structure (sorus) which is full of black mass of powdery spores in place of grains. The smut sorus are also seen on stems and axils of leaves

Causal organism: *Ustilago crusgalli*

Management

i) Growing of resistant varieties

ii) Use of certified seeds

iii) Seed treatment with Carboxin 37.5% WP +Thiram 37.5% WP @3gm/kg of seed

Minor diseases

Fungal diseases

Ergot *Claviceps microcephala*

Leaf spot *Colletotrichum graminicolum*

Leaf spot *Curvularia lunata*

Leaf spot *Helminthosporium frumentacei*

Leaf spot *Helminthosporium setariae*

Shoot smut *Ustilago crusgalli*

COMMON MILLET (*Panicum miliaceaum*)

Fungal diseases

1) Head Smut

Symptoms: The entire inflorescence become smutted gall like structure which is covered by grayish white membrane. The smutted galls contaminate healthy grains as the membrane rupture before harvest.

Causal organism: *Sphacelotheca destruens*

Management

i) Rogue out of smutted ears and destroyed them.

ii) Growing of resistant varieties.

iii) Use of certified seeds.

iv) Soaking the seeds in cold water from 6AM to 10AM on a bright sunny day (within the period of May-15 to June-15) and exposing them to sun on a cemented floor from 10AM to 5PM.

v) Seed treatment with Carboxin 37.5% WP +Thiram 37.5% WP @ 3gm/kg of seed

2. Downy mildew

Symptoms: Downy growth of the fungus on the leaves and leaf shredding occur in advanced stage of growth. Green ear symptoms develop on rare cases.

Causal organism: *Sclerospora graminicola*

Management

i) Field sanitation.

ii) Growing of resistant varieties

iii) Use of certified seeds

iv) Seed treatment with Carboxin 37.5% WP +Thiram 37.5% WP @3gm/kg of seed

Bacterial diseases

1) Bacterial streak

Symptoms: Longitudinal water-soaked streaks usally 2 to 15 m long and 2-3 mm wide streaks are developed on the infected leaves.

Causal organism: *Xanthomonas vasicola pv. holcicola*

Management

i) Field sanitation. Rogue out of diseased plants immediately for the infection

ii) Growing of resistant varieties

iii) Use of certified seeds

iv) Seed treatment with 5% Magnesium arsenate @1g/kg of seed

Minor diseases

Fungal diseases

Leaf spot: *Drechslera oryzae*

Leaf spot: *Helminthosporium panici-miliacei*

Blast: *Pyricularia grisea*

Leaf rust: *Uromyces linearis*

3

Diseases of Pulses

GREEN GRAM (*Vigna radiata*) & BLACK GRAM (*Vigna mungo*)

Fungal diseases

1) Seed and seedling rot

Symptoms: The seeds rot before germination and after germination the seedlings also show rotting. The disease can be easily identified by poor germination and poor stand of the crop.

Causal organisms

i) *Fusarium* sp

ii) *Pythium* sp

iii) *Rhizoctonia solani*

iv) *Macrophomina phaseoli*

Management

i) Field sanitation

ii) Use of certified seeds or sowing of seeds collected from healthy plants of disease free field.

iii) Seed treatment with Thiram 75%WS @3g/kg or Carboxin 37.5% WP+Thiram 37.5% WP @3gm/kg of seed.

2. Charcoal rot

Symptoms: The rotting starts from the roots and progress towards the stem where reddish brown to black coloured lesions are formed near the soil level and the affected stems turn black. Plants usually show rotting one month after sowing and the affected plants may die.

Causal organism: *Macrophomina phaseolina*

Management

i) Field sanitation

ii) Use of certified seeds or sowing of seeds collected from healthy plants of disease free field.

iii) Seed treatment with Carboxin 37.5% WP +Thiram 37.5% WP (3g/kg) of seed.

iv) Follow the crop rotation with jowar or bajra for three years

3) Cercospora leaf spot

Symptoms: Small, round or angular brown spots develop on the leaves. The centre is ashy coloured with reddish purple border. Leaves,stalks and pods are affected. In severe infection the leaves may dry and the affected pods become blackened.

Causal organism: *Cercospora cruenta* and *C. dolichii*

Management

i) Field sanitation

ii) Use of certified seeds or sowing of seeds collected from healthy plants of disease free field.

iii) Seed treatment with Carboxin 37.5% WP +Thiram 37.5% WP (3g/kg) of seed.

iv) Spraying the crop with Carbendazim (12%)+ Mancozeb (63%W. P) @0.3%. If necessary, spraying is recommended for twice /thrice at 10 days interval.

v) Growing of some regional resistant varieties like Sashi, Kamadev, Durga and OBGG-32 in green gram and LPG-787 and PU-31 in black gram.

4) Anthracnose

Symptoms: Brown circular spots appear on leaves. These spots increases in size become irregular dark lesions. On the pods black sunken lesions may be found.

Causal organism: *Glomerella lindemuthianum*

Management

i) Field sanitation

ii) Use of certified seeds or sowing of seeds collected from healthy plants of disease free field.

i) Seed treatment with Carboxin 37.5% WP +Thiram 37.5% WP @3gm/kg of seed.

ii) Spraying the crop with Carbendazim (12%)+ Mancozeb (63%W.P) @ 0.3%. If necessary spraying is recommended for twice /thrice at 7 days interval.

iii) Growing of some regional resistant varieties like Sashi, Kamadev,Durga and OBGG-32 in green gram

5) Powdery mildew

Symptoms: White powdery patches appear on the leaves and other green parts. Later on, these patches become dull coloured and are covered with black dot like cleistothecia

Causal organism: *Erysiphe polygoni*

Management

i) Field sanitation

ii) Use of certified seeds or sowing of seeds collected from healthy plants of disease free field. .

iii) Spray the crop with Sulphur 80%WP @ 4.0g/litre

6) Ascochyta blight

Symptoms: Large brown colour spots appear on the leaves. Later black dot like structures (fruiting bodies) appear on the necrotic lesions.

Causal organism: *Ascochyta phaseolorum*

Management

i) Field sanitation

ii) Use of certified seeds or sowing of seeds collected from healthy plants of disease free field. .

iii) Spray the crop with Carbendazim (12%)+ Mancozeb (63%W.P) @ 0.3%. If necessary, spraying is recommended for twice or thrice at 7 days interval

Viral diseases

1) Leaf curl

Symptoms: The infected leaves show curling of margins downwards. The veins on the under surface of leaves show reddish brown discolouration. The growth of the plants stopped and plants become stunted

Causal organism: Leaf curl virus

Management

i) Field sanitation

ii) Use of certified seeds or sowing of seeds collected from healthy plants of disease free field.

iii) Roguing out of diseased plants from the field. Removal of weeds

iv) Spray Metasystox (0.1%) for controlling the vectors. Two or three sprayings are recommended at 10 days interval.

2) Leaf crinkle

Symptoms: The symptoms develop after 3 to 4 weeks of sowing. Leaves become thicker and leathery

Causal organism: Leaf crinkle virus

Management

i) Sowing of healthy seeds.

ii) Roguing out of diseased plants

iii) Spray Metasystox (0.1%) for controlling the vectors. Two or three sprayings are recommended at 10 days interval.

3) Yellow mosaic

Symptoms: Initially yellow round spots appear on the leaf lamina within one month after sowing. These spots enlarges and the leaves show yellow patches alternating with green patches. The affected leaves become completely yellow and get reduced in size. In later stage the completey yellow leaves emerge from the plants. The pods become yellow in severe cases. Sometimes there may be no yield.

Causal organism: Yellow mosaic virus

Management

i) Field sanitation

ii) Use of certified seeds or sowing of seeds collected from healthy plants of disease free field.

iii) Growing of regional resistant varieties like IPM-2-3, IPM-0-2-14, Samrat, PDM-139 in green gram, PU-31 and PU-35 in blackgram.

iv) Spraying the crop with Imidacloprid (0.025%) to check the vector population.

Fungal diseases

4) Mosaic mottle

Symptoms: Appearance of irregular light green areas alternating with normal green areas. The leaves emerge after infection become smaller in size and margins show upward rolling. In later stage there is raising of interveinal areas.

Causal organism: Mosaic mottle virus

Vector: Aphids

Management

i) Field sanitation. Removal of weeds from the field.

ii) Use of certified seeds

iii) Roguing out of the diseased plants immediately after appearance of the disease in the field to prevent further spread of the disease.

iv) Spray the crop with Metasystox 25 EC (0.1%) for controlling the vectors. Two or three sprayings are recommended at 10 days interval

Minor diseases

Fungal diseases

Seed infection *Colletotrichum truncatum*

Seedling blight *Phomopsis* sp

Leaf spot *Corynespora cassicola*

Leaf spot *Septoria leguminose*

Leaf spot *Cercospora kikuchii*

Leaf spot *Colletotrichum demantium*

Leaf spot and blight *Phyllosticta phaseolina*

Stem rot *Sclerotinia sclerotiorum*

Charcoal rot *Pellicularia filamentosa*

Leaf and stem gall *Synchytrium ajerkari*

Leaf and stem gall *Synchytrium phaseoli*

Leaf and stem gall *Synchytrium phaseoli-radiati*

Viral diseases

Mosaic: Bean common mosaic virus (BCMV)
Cowpea mosaic virus (CPMV)

PIGEON PEA (*Cajanus cajan*)

1) Wilt

Symptoms: The disease is observed both in seedling and adult plants. Wilting of plant is gradual. Yellowing, withering and drying of leaves are found followed by drying of entire plant. The roots become dark brown and show black streak below the bark. Root lets are damaged. Wilting of plants occur in patches in the field.

Causal organism: *Fusarium oxysporum* f. sp. *udum.*

Management

i) Crop rotation for four years.

ii) Mixed cropping of arhar with jowar is recommended in heavily infested area.

iii) Collection and removal of diseased crop debris and deeply buried them in soil.

iv) Growing of resistant varieties

v) Seed treatment with Tebuconazole 25% DS @2g/kg

vi) Basal drenching with Carbendazim 50% WP @2g /litre

2. Stem Rot

Symptoms: In this disease there is development of brown to dark brown lesions on the stems near the soil surface. These lesions girdle the whole stem and the plant dies.

Causal organism: *Phytophthora drechslera* var. *cajani*

Management

i) Clean cultivation

ii) Growing of resistant varieties

iii) Basal drenching with Metalaxyl 8% +Mancozeb 64% WP @ 2.0 g/litre

3) Leaf spot

Symptoms: Small brown to black spots appear on leaves. These spots enlarge in size and the centre becomes ashy colour. Several spots coalesce cause blightening of leaves. Sometimes elongated brown to black lesions are found on stems and petioles

Causal organism: *Cercospora indica*

Management

i) Clean cultivation

ii) Crop rotation for two years.

iii) Growing of resistant varieties

iv) Spraying the crop with Carbendazim 12% + Mancozeb 63% W.P @0.3% twice /thrice at 7 days interval.

4) Powdery mildew

Symptoms: Powdery patches appear on both the surface of leaves and other green parts. Later on, these patches become dull coloured are covered with black dot like cleistothecia. The leaves become dry and fall down.

Causal organism: *Erysiphe polygoni*

Management

i) Clean cultivation. Removal of crop debris after harvest of the crop and buried deeply in soil.

ii) Removal of weeds.

iii) Growing of resistant varieties

iv) Spraying of crop with Sulphur 80% WP @4.0 g/litre twice or thrice at seven days interval.

5) Stem Canker

Symptoms: Grey coloured elliptical lesions are found on the collar region of the plant. These lesions convert into deep cankers. These cankers girdle the stem and the plants collapse.

Causal organism

i) *Diplodia cajani*

ii) *Colletotrichum cajani*

Management

i) Avoid injury during intercultural operation

ii) Crop rotation

iii) Basal drenching of the crop with Carbendazim 12% + Mancozeb 63% W.P @0.3% twice /thrice at 7 days interval.

Bacterial diseases

1) Bacterial leaf spot

Symptoms: Minute circular water-soaked spots develop on leaves. These spots become rectangular and surrounded by chlorotic halo. Older spots become dark brown. These spots coalesce with each other to form larger patches. The leaves are distorted and fall down.

Causal organism: *Xanthomonas cajani*

Management

i) Clean cultivation

ii) Growing of resistant varieties

iii) Spraying of crop with Copper oxychloride 50% WP @ 2. 5 g/litre

Viral diseases

1) Sterility Mosaic

Symptoms: The affected plants become stunted in growth and branch profusely. Mosaic symptoms are observed on leaves. In case of mild infection the plants may bear flowers but the pod production is very less. In severe infection the plants remain sterile (no flower and no pod production). In tolerant varieties a few branches in plants are affected while other parts remain healthy.

Causal organism: Sterility mosaic virus.

Vector: Eriophyde mite. (*Aceria cajani*)

Management

i) Clean cultivation

ii) Rouging out of infected plant upto 40 days after sowing.

iii) Growing of resistant varieties

iv) Spraying of crop with Fenazaquin @ 1ml/litre immediately after appearance of the disease in the field and if necessary second spraying is applied at an interval of 15 days for controlling mites.

Minor diseases

Fungal diseases

Collar rot *Pythium aphanidermatum*

Leaf spot *Cercospora cajani*

Leaf spot *Alternaria tenuissima*

Leaf spot *Cercospora cajanicola*

Leaf and stem spot *Cercospora indica*

Leaf spot *Phyllosticta cajani*

Stem cankers *Cercospora cajani*

Leaf rust *Uredo cajani*

Leaf gall *Synchytrium phaseoli-radiata*

Stem rot *Rhizoctonia solani*

Root and stem rot *Sclerotium rolfsii*

Stem canker *Macrophomina cajanicola*

Phoma canker *Phoma* sp.

Viral diseases

Mosaic Mungbean yellow mosaic virus (MYMV)

Yellow mosaic virus (YMV)

CHICKPEA (*Cicer arietinum*)

1) Root Rot

Symptoms: Both the the disease can attack at any stage of plant growth (from seedling to matured stage). The leaves become pale green and later turn yellow and ultimately dry out. Finally the whole plant look yellowish followed by drying from top downwards.

Causal organism: *Opercullela padwickii*

Management

i) Clean cultivation

ii) Crop rotation

2) Sclerotinia blight

Symptoms: All the above ground parts of the plant are affected. The affected plants become yellow then brown ultimately die. White cottony growth of the fungus along with small hard colour sclerotia may be seen on the stem. The spots girdle the stem causing drying.

Causal organism: *Sclerotinia sclerotiorum*

Management

i) Clean cultivation

ii) Crop rotation for two years

iii) Growing of resistant varieties

3) Wilt

Symptoms: The disease can attack at any stage of plant growth (from seedling to matured stage). The leaves become pale yellow and ultimately dry out.

Roots turn black. Finally the whole plant look yellowish followed by drying

Causal organism: *Fusarium oxysporum* f. sp. *ciceri*

Management

i) Clean cultivation

ii) Crop rotation for two years.

iii) Growing of resistant varieties like G-24 or C-214.

4) Grey mold

Symptoms: Production of necrotic spots on leaves, twigs, petioles and flowers. The branches and stems are also affected. The fungus grow rapidly in humid weather. The affected stem finally breaks up and the plant dies out.

Causal organism: *Botrytis cinerea*

Management

i) Late sown crop escapes the disease infection

ii) Spraying of Mancozeb 75%WP (3g/litre)

5) Rust

Symptoms: The pustules appear first on under surface then in upper surface of leaves. The affected leaves dries up.

Causal organism: *Uromyces ciceris arietini*

Management

i) Seed treatment with Carboxin 37.5% WP +Thiram 37.5% WP @ 3g/kg of seeds

ii) Growing of resistant varieties

iii) Spraying of Mancozeb 75%WP (3g/litre) twice or thrice at 7 days interval.

6) Powdery mildew

Symptoms: White powdery growth appear on leaves and spread to cover the stem and other plant parts. The infection may occur at any stage of plant growth but become severe when the plants are in flowering stage. The entire plant dries.

Causal organism: *Erysiphe* sp

Management

i) Field sanitation

ii) Use of certified seeds or sowing of seeds collected from healthy plants of disease free field. .

iii) Spray the crop with any Wettable Sulphur 80% WG @ 2.5g/litre

7) Ascochyta blight

Symptoms: The disease appears as brown spots on stems, leaves and pods which become straw coloured and in the centre dot like dark brown pycnidia are present which are arranged in circles.

Causal organism: *Ascochyta rabiei*

Management

i) Seed treatment with Carboxin 37.5% WP +Thiram 37.5% WP @ 3g/kg of seeds.

ii) Growing of resistant varieties.

iii) Spraying of Mancozeb 75% WP(3g/litre) after apperance of the disease in the field

Minor diseases

Fungal diseases

Leaf and stem blight: *Mycospharella rabiei*

Powdery mildew: *Erysiphe* sp

Stem and root rot: *Pellicularia rolfsii*

Sclerotinia blight: *Sclerotinia sclerotiorum*

Verticillium wilt: *Verticillium albo-atrum*

Viral diseases

Mosaic

Bean yellow mosaic virus (BYMV)

Cucumber mosaic virus (CMV)

Leaf rolling: Pea leaf roll virus (PLRV)

COWPEA (*Vigna sinensis*)

1) Root and stem rot

Symptoms: Rotting of root and stem of the plant at seedling stage.

Causal organism: *Pythium aphanidermatum*

Management

i) Seed treatment with Carboxin 37.5% WP + Thiram 37.5% WP /kg of seeds

ii) Growing of resistant varieties

2) Powdery mildew

Symptoms: White powdery growth appear on leaves and spread to cover the stem and other plant parts. The infection may occur at any stage of plant growth but become severe when the plants are in flowering stage. The entire plant dries.

Causal organism: *Erysiphe polygoni*

Management

i) Field sanitation

ii) Use of certified seeds or sowing of seeds collected from healthy plants of disease free field.

iii) Spray the crop with any Wettable Sulphur 80% WG@2. 5g/litre

3) Anthracnose

Symptoms: Dark brown sunken spots with reddish brown margin appear on the leaves. In moist weather black fruiting bodies appear on these spots. In severe condition fruits and stems are affected.

Causal organism: *Glomerella lindemuthianum*

Management

i) Seed treatment with Carboxin 37.5% WP +Thiram 37.5% WP @ 3g/kg of seeds

ii) Growing of resistant varieties

iii) Spraying of Mancozeb 50%WS (3g/litre) or Carbendazim 50%WP twice or thrice at 7 days interval.

4) Rust

Symptoms: Numerous brown pustules develop on the under surface of the leaves. Sometimes pustules develop on stems and pods. At the end of the season the uredospores may be turned into black.

Causal organism: *Uromyces appendiculatus*

Management

i) Spraying of crop with Propiconazole 50% WP @ 1.5 gm /litre

ii) Growing of resistant varieties

5) Die back

Symptoms: The twigs and branches dry up from the tip down wards. Small black dot like acervuli of the fungus appear on the dried up portions. Infected pods become shrivelled.

Causal organism: *Colletotrichum capsici*

Management

i) Field sanitation

ii) Crop rotation

iii) Spraying of crop with Mancozeb 75% WP @3 gm/litre two to three times at 7 days interval

iv) Growing of resistant varieties

Bacterial Diseases

Bacterial Blight

Symptoms: The cotyledons become red and shrivelled. The margin of trifoliate leaves become necrotic. Cankers develop on stems. Growing points are damaged. Due to presence of cankers on the stem the plants may break at the time of high wind speed.

Causal organism: *Xanthomonas axonopodis* pv. *vignicola*

Management

i) Clean cultivation

ii) Growing of resistant varieties

iii) Spraying of crop with Copper oxychloride 50% WP @ 2.5 g/litre

Viral diseases

1) Cowpea mosaic

Symptoms: The affected leaves become pale yellow and exhibit mosaic and vein banding symptoms. The size of the affected leaves is reduced and show puckering. Number of pods are reduced and deformed. Ultimately the whole plants become yellow.

Causal organism: Cowpea mosic virus (Como Virus group)

Vector: aphids.

Management

i) Use of healthy seeds (Certified seeds)

ii) Spraying of crop with Metasystox (0.1%) /litre

ii) Growing of resistant varieties

Minor diseases:

Fungal diseases

Leaf spot: *Alternaria alternata*

Leaf spot: *Ascochyta phaseolorum*

Leaf spot: *Cercospora cruenta*

Leaf spot: *Cercospora canescens*

Leaf spot: *Cladosporium vignae*

Leaf spot: *Corynespora cassiicola*

Leaf spot: *Leptosphaerulina trifolii*

Leaf blight: *Phytophthora vignea*

Southern blight: *Botryobasidium rolfsii*

Stem canker and web blight: *Rhizoctonia solani*

Root rot: *Sclerotium rolfsii*

Rust: *Uromyces phaseoli*

Ashy stem blight: *Macrophomina phaseolina*

Tip burn *Alternaria tenuis*

Ozonium wilt *Ozonium texanum*

Verticillium wilt *Verticillium albo-atrum*

Dry root rot *Macrophomina phaseoli*

Pod rot *Choanephora cucurbitarum*

Bacterial diseases

Bacterial leaf spot *Pseudomonas syringae*

Bacterial wilt *Pseudomonas solanacearum*

Viral diseases

Mosaic

Cucumber mosaic virus (CMV)

Southern bean mosaic virus (SBMV)

Tobacco mosaic virus (TMV)

PEA (*Pisum sativum*)

Fungal Diseases

1) Seed and root rot

Symptoms: Rotting of seeds occur before germination inside the soil. Rotting of germinated seeds and rotting of root portion of young seedlings occur.

Causal organism: *P. aphanidermatum*

Management

i) Seed treatment with Carboxin 37.5% WP +Thiram 37.5% WP @3gm/kg of seed

ii) Soil treatment

2) Fusarium Root Rot

Symptoms: Reddish brown sunken lesions appear on the tap root. Hence the roots rot and the plants show yellowing of lower leaves and gradually the plants wilt.

Causal organism: *Fusarium solani* f. sp. *pisi*

Management

i) Seed treatment with Carboxin 37.5% WP +Thiram 37.5% WP @3gm/kg of seed

ii) Growing of resistant varieties

iii) Follow the crop rotation for 3 to 4 years with non-host crops

3) Fusarium wilt

Symptoms: At the time of flowering the lower leaves become yellow. The growth of the plant is arrested and there is downward curling of leaves and finally the entire plant wilts.

Causal organism: *Fusarium oxysporum* f.sp *pisi*

Management

i) Avoid early sowing

ii) Seed treatment with Carboxin 37.5% WP +Thiram 37.5% WP @3g/kg of seeds

ii) Growing of resistant varieties

iii) Follow the crop rotation for 3 to 4 years with non-host crops

4) Downy mildew

Symptoms: On the upper surface of the leaves scattered yellow to irregular shape of brown patches appear.The corresponding lower surface shows whitish to greyish violet downy growth of the fungus

Causal organism: *Peronospora pisi*

Management

i) Crop rotation

ii) Field sanitation

iii) Spraying the crop with Mancozeb 75%WP @ 3 gm/litre two to three times at 7 days interval.

5) Powdery mildew

Symptoms: Whitish to greyish powdery patches appear on the upper surface and then on the under surface of leaves. In severe infection stems, tendrils and pods are infected. At the later stage the powdery growth become dark grey and finally black on which numerous small round dark cleistothecia develop.

Causal organism: *Erysiphe polygoni*

Management

i) Field sanitation. ii) Spray the crop with Wettable sulphur 80% WG@2.5g/ litre.

Bacterial diseases

1) Bacterial Blight

Symptoms: Water soaked lesions develop on all the above ground parts like leaves, stems and pods. On the leaf lesions turn brown and glistening.

Causal organism: *Pseudomonas pisi*

Management

Crop rotation

Minor diseases

Fungal diseases

Anthracnose *Colletotrichum gloeosporioides*

Anthracnose *Colletotrichum pisi*

Ascochyta blight *Ascochyta pinodella*

Collar rot *Pythium* sp

Leaf blight *Ascochyta .pisi*

Leaf spot *Alternaria alternata*

Leaf spot *Cercospora pisa-sativae*

Leaf spot *Phyllosticta pisi*

Leaf spot *Septoria pisi*

Root rot *Aphanomyces euteiches*

Root rot *Rhizoctonia bataticola*

Root rot and stem rot *Sclerotium rolfsii*

Bacterial diseases

Bacterial blight *Pseudomonas pisi*

Viral diseases

Mosaic

Pea seed-borne mosaic virus (PSbMV)

Pea top necrosis virus (PTNV)

HORSE GRAM (*Dolichos biflorus*)

Fungal diseases

1) Dry root rot

Symptoms: The rotting starts from the roots and progress towards the stem where brown to black coloured lesions are formed near the soil level and the affected stems turn black.

Causal organism: *Macrophomina phaseoli*

Management

i) Seed treatment with Carboxin 37.5% WP +Thiram 37.5% WP @3gm/kg of seed

ii) Growing of resistant varieties

2) Root rot

Symptoms: Rotting of root portion of the plants occur.

Causal organism: *Pellicularia filamentosa*

Management

i) Seed treatment with Carboxin 37.5% WP +Thiram 37.5/WP @ 3gm/kg of seed

ii) Growing of resistant varieties

3) Anthracnose

Symptoms: Dark brown circular spots appear on leaves. These spots increase in size become irregular dark lesions. On the pods black sunken lesions are found.

Causal organism: *Glomerella lindemuthianum*

Management

i) Seed treatment with Carboxin 37.5% WP +Thiram 37.5%WP @3gm/kg of seed

ii) Growing of resistant varieties i.e. Urmi

iii) Spraying of Mancozeb 75% WP (3g/litre)

4) Leaf spot

Symptoms: There are light brown spots on the leaves. These spots coalesce with each other which increase in size to form bigger patches on the leaves. Severe leaf infection cause defoliation of the plants.

Causal organism: *Cercospora dolichi*

Control measures

i) Seed treatment with Carboxin 37.5% WP +Thiram 37.5% WP @3g/kg of seeds

ii) Growing of tolerant varieties,i.e.Urmi

iii) Spraying of Mancozeb 50%WS (3g/litre) after appearance of the diseases in the field

5) Die back

Symptoms: The twigs and branches dry up from the tip downwards. Small black dot like acervuli of the fungus appear on the dried up portions. Infected pods shrivel.

Causal organism: *Colletotrichum capsici*

Management

i) Seed treatment with Carboxin 37.5% WP @ Thiram 37.5% WP @3g/kg of seeds

ii) Growing of resistant varieties

iii) Spraying of Mancozeb 50% WS (3g/litre) after appliance of the diseases in the field

Minor diseases

Fungal diseases

Leaf spot: *Alternaria tenuisima*

Leaf spot: *Pyrenochaeta dolichi*

Leaf spot: *Phyllosticta dolichi*

Wilt: *Fusarium oxysporum*

Rust: *Uromyces appendiculatus*

Viral diseases

Mosaic: Mung bean yellow mosaic virus (MYMV)

SOYBEAN (*Glycine max*)

Fungal diseases

1) Damping-off

There are two types of damping-off.

i) Pre-emergence

ii) Post-emergence

Symptoms: In pre-emergence damping-off the germinating seeds are very often covered with fungal growth.

In post-emergence damping off the collar region of the stem in the affected seedlings turn brown. Roots become brown. These plants show wilting,the leaves become grey-green and then turn brown.

Causal organism: *Pythium ultimum*

Management

i) Seed treatment with Carbendazim 50% WP @ 2. 5 g per kg of seed.

ii) Improve the drainage facility

2) Root rot and Tip blight

Root rot

Symptoms: The affected plants suddenly wilt and die. The bark of the stem, the primary and secondary roots are shredded and covered with sclerotia. Pods in the affected plants are partially filled up.

Tip blight

Symptoms: On tender leaves reddish brown spots appear and the vein become dark brown. The leaves wither and dry up. The disease spreads to the stem through petiole. The entire plant die within 10 days of attack. In affected plants less number of pods are formed with fewer number of seeds.

Causal organism: *Rhizoctonia solani*

Management

i) Field sanitation. Diseased plants should be removed and destroyed.

ii) Seed treatment with Carboxin 37.5% WP +Thiram 37.5% WP @ 3g/kg of seeds

iii) Growing of resistant varieties

iv) Basal drenching with Metalaxyl 8% + Mancozeb 64% WP @ (0.2%)

3) Collar Rot

Symptoms: Water-soaked cream coloured lesions develop on cotyledons and hypocotyls which enlarge longitudinally on the hypocotyl to the radicle causing death of the seedlings. Yellowing of leaves occur prematurely and the plants die. When a diseased plant is split open longitudinally the vascular tissues look brown or blackish.

Causal organism: *Fusarium oxysporum* f. sp. *tracheiphilum*

Management

i) Clean cultivation

ii) Crop rotation for two years.

iii) Growing of resistant varieties

4) Cercospora leaf spot

Symptoms: Small circular to angular reddish-brown spots appear on the foliage. The spots increase in size and develop brown to white centre with reddish brown border. Several spots may coalesce to form large irregular areas. In severe infection on stem elongated reddish lesions develop with black border. On pods lesions are circular brown to grey with a narrow dark brown ring.

Causal organism: *Cercospora sojina*

Management

i) Clean cultivation

ii) Crop rotation for two years.

iii) Growing of resistant varieties

5) Purple stain disease

Symptoms: Large irregular reddish brown spots appear on leaves. In later stage of growth reddish brown spots appear on stems and pods.

Causal organism: *Cercospora kikuchii*

Management

i) Seed treatment with Carboxin 37.5% WP +Thiram 37.5% WP @3g/kg of seeds

ii) Growing of resistant varieties

iii) Spraying of Mancozeb 50% WS (3g/litre)

6) Septoria leaf spot

Symptoms: Small brown spots from pin head size to 5 mm diameter appear on the leaves. In severe attack the leaves prematurely fall. The stem,petiole and pods are also affected on which brownish lesions of different shape appear and pycnidia develop at later stage.

Causal organism: *Septoria glycines*

Management

i) Field sanitation

ii) Use of certified seeds or sowing of seeds collected from healthy plants of disease free field.

iii) Seed treatment with Carboxin 37.5% WP +Thiram 37.5% WP (0.3%)

iv) Spray the crop with Carbendazim (12% WP)+ Mancozeb (63% WP) @ 0.3%. If necessary the spraying is recommended for twice /thrice at 7 days interval

7) Downy Mildew

Symptoms: Yellow green areas appear on the upper surface of the leaves. As the disease advances the yellowish green areas become greyish brown. A greyish mold growth is observed on the corresponding under surface of the lesion. Severly infected leaves fall prematurely.

Causal organism: *Peronospora manshurica*

Management

i) Field sanitation

ii) Use of certified seeds or sowing of seeds collected from healthy plants of disease free field.

iii) Growing of resistant varieties.

iv) Spray the crop with Carbendazim (12%WP)+ Mancozeb (63%W. P) @0.3%. If necessary spraying is recommended for twice or thrice at 7 days interval

8) Sclerotium blight or Southern stem blight

Symptoms: Large brownish round sclerotia are observed on stem. Infected plants die prematurely may be before pod formation

Causal organism: *Sclerotium* sp

Management

i) Field sanitation

ii) Deep summer ploughing. Infected soybean fields should be ploughed up to depth of 12-15cm.

iii) Use of certified seeds or sowing of seeds collected from healthy plants of disease free field.

iv) Growing of resistant varieties.

v) Spraying of Mancozeb 50% WS (3g/litre) during the period of flowering 2 to 3 times at 15 days interval.

9) Drechslera blight

Symptoms: Brown circular and angular spots appear on the leaves. As the spots enlarge the centre become grey with dark margins. Several spots coalesce cause blightening of leaves

Causal organism: *Drechslera glycine*

Management

i) Seed treatment with Vitavax Power @3g/kg of seeds

ii) Growing of resistant varieties

iii) Spraying of Mancozeb 50% WS(3g/litre)

10) Pod and stem blight

Symptoms: Light brown spots after on stems and pods. The pycnidia appear on these spots in the form of small dots. There may be no seed formation if formed the seeds become shrivelled.

Causal organism: *Diaporthe phaseolorum*

Management

i) Use of certified seeds or sowing of seeds collected from healthy plants of disease free field.

ii) Seed treatment with Carboxin 37.5% WP +Thiram 37.5% WP @3gm/kg of seed.

iii) Growing of resistant varieties

iv) Spraying of Mancozeb(3g/litre)

11) Target spot

Symptoms: The fungus produces reddish brown circular spot, i.e. tiny dots to 1 cm diameter. The concentric rings develop on larger spots. The central dead area is surrounded by yellow green halo.

Causal organism: *Corynespora cassiicola*

Management

i) Seed treatment with Carboxin 37.5% WP +Thiram 37.5% WP @3g/kg of seeds

ii) Growing of resistant varieties

iii) Spraying of Mancozeb 50% WS(3g/litre)

12) Stem canker

Symptoms: The brown sunken lesion girdles the stem, base of the petiole is affected. Tiny black dots appear on dead stems.

Causal organism: *Diaporthe phaseolorum*

Management

i) Seed treatment with Vitavax Power @3g/kg of seeds

ii) Growing of resistant varieties

iii) Spraying of Mancozeb 50% WS (3g/litre)

13) Anthracnose

Symptoms: The fungus produces lesions on petioles,stems and pods. The pods become yellowish green and dry. The seeds in diseased pods may be shrivelled and covered with fungal mycelium.

Causal organism: *Colletotrichum truncatum*

Management

i) Seed treatment with Vitavax Power @3g/kg of seeds

ii) Growing of resistant varieties

iii) Spraying of Mancozeb 50%WS(3g/litre)

14) Powdery mildew

Symptoms: Whitish to greyish powdery patches appear on the upper surface of leaves. In severe infection stems and pods are infected. At the later stage the infected leaves dry followed by defoliation.

Causal organism: *Microsphaera diffusa*

Management

i) Seed treatment with Carboxin 37.5% WP +Thiram 37.5% WP @3g/kg of seeds

ii) Growing of resistant varieties

iii) Spraying of Sulfex or Karathane (2g/litre)

15) Rust

Symptoms: Brownish pustules develop on entire surface of leaves. Pustules rupture and expose dry powdery mass of spores. Untimely defoliation of leaves occur. Number, size and weight of the pods reduce.

Causal organism: *Phakospora pachyrhizi*

Management

i) Seed treatment with Vitavax Power @3g/kg of seeds.

ii) Growing of resistant varieties

iii) Spraying of Mancozeb 50% WS(3g/litre)

Bacterial diseases

1) Bacterial Pustule

Symptoms: Small pale green to reddish brown pustules appear on both the surface of the leaves. The central portion appears slightly raised. Mature spots become angular and dark brown. Several spots may develop on each leaf which may coalesce to form larger patches. There may be defoliation. Due to heavy infestation. Minute brown spots may appear on the pods.

Causal organism: *Xanthomonas phaseoli* var. *sojensis.*

Management

i) Use of seeds from disease free plant.

ii) Growing of resistant varieties.

2) Wild fire

Symptoms: Brown lesions appear on the leaves. In humid weather the lesions enlarge and form large brown dead area. Yellow holes are formed on dead areas of the leaf.

Causal organism: *Pseudomonas syringae* pv *tabaci*

Management

i) Diseased plants should be carefully removed and burried deep in soil.

ii) Growing of resistant varieties

iii) Bacterial blight

Bacterial Blight

Symptoms: The disease appear on the leaves in form of small yellow spots surrounded by yellow green halo and turn dark brown to black. Bacterial ooze is discharged from the leaves which dry in air and form silvery films. Several spots coalesce form larger patch. Same symptoms arise on stems and pods. Severe infections result in premature shedding of the leaf.

Causal organism: *Pseudomons syringae* pv.*glycinea*

Management

i) Diseased plants should be carefully removed and burried deep in soil.

ii) Growing of resistant varieties

iii) Use of certified seeds

iv) Seed treatment with Carboxin 37.5% WP +Thiram 37.5% WP @3g/kg of seeds

v) Spraying of Mancozeb 50% WP (3g/litre)

Viral Diseases

1) Yellow mosaic

Symptoms: The leaves become bright yellow. In some cases diffuse patches appear on leaves which gradually increase in size and coalesce the entire lamina become yellow. The size of the leaflets reduce and become narrower. There is reduction in number and size of pods in infected plant. If plants are infected 75 days before planting there is much reduction in yield of the plant.

Causal organism: Mungbean yellow mosaic virus (MYMV)

Vector: White fly

Management

i) Field sanitation

ii) Growing of resistant varieties

iii) Use of certified seeds

iv) Removal of weeds especially leguminous weeds

v) Rouging of infected plants should be regular practice

vi) Spraying of crop with Metasystox (0.1%) at 7 days interval to check the white fly population.

2) Soybean Mosaic

Symptoms: The unifoliate leaves of plants developed from infected seeds become chlorotic and the leaves show rolling. There is reduction in size of trifoliate leaves. These leaves become mottled and rugose type. In same season infection of young leaves show vein clearing, mottling and rugosity. The petioles and internodes shortened Seed production is delayed and reduced.

Causal organism: Soja virus 1

Management

i) Field sanitation

ii) Growing of resistant varieties

iii) Use of certified seeds

iv) Removal of weeds especially leguminous weeds

v) Rouging of infected plants should be regular practice

vi) Spraying of crop with Imidacloprid 17.8%SL@1.0 ml/3 litre or Thiomethoxam 25% WG@1.0g/5 litre at 7 days interval to check the white fly population.

3) Bud Blight

Symptoms: The seedlings develop from infected seeds show browning of top leaves which curve downward become dry and brittle. When the infection occur before flowering the apical bud and shoot become brown and curve downward and become brittle. There may be excessive proliferation of leaves, floral buds and production of under sized pod which may or may not contain seed.

Causal organism: Tobacco ring spot virus

Vector: Thrips (*Frankliniela schultzei, Scirtothrips dorsalis)*

Management

i) Field sanitation

ii) Use of certified seeds or sowing of seeds collected from healthy plants of disease free field. .

iii) Spraying of Imidacloprid 17.8%SL@1.0 ml/3 litre or Thiomethoxam 25% WG@1.0g/5 litre

Phytoplasma diseases

Phyllody

Symptoms: The infected plants are stunted and thinner. Clusters of leafy structures develop from the leaf axils which make the plant a bushy like appearance. The floral parts are converted into green leafy structures.

Causal organism: Mycoplasma like organism

Management

i) Field sanitation

ii) Use of certified seeds or sowing of seeds collected from healthy plants of disease free field. .

iii) Spraying of Imidacloprid 17.8%SL@1.0 ml/3 litre or Thiomethoxam 25% WG@1.0g/5 litre

Nematode diseases

1) Root knot nematode

Symptoms: Infected plants become stunted and leaves turn yellow. Plants show wilting during hot and dry weather. Galls of different sizes are present on the roots which obstructs the flow of water and nutrients to leaves.

Causal organism

i) *Meloidogyne incognita*

ii) *M. javanica*

Management

i) Crop rotation with non-host crops like maize and cotton for three years.

ii) Growing of resistant varieties

2) Soybean cyst nematode

Symptoms: Infected plants remain stunted, chlorotic and leaves shed prematurely. Severely infected plants have necrotic root system. Rhizobium nodule formation in plants may be affected. The number of nodules per plant are reduced. In severe infection there may be no nodule formation.

Causal organism: *Heterodera glycines*

Management

i) Crop rotation with non-host crops like maize and cotton for three years.

ii) Growing of resistant varieties

Minor diseases

Fungal diseases

Seedling rot *Nectaria haematococca*

Seedling rot *Fusarium graminearum*

Leaf spot *Phyllosticta sojaecola*

Leaf spot *Phyllosticta glycinate*

Leaf spot *Stemphylium botryosum*

Leaf spot *Cercospora diazii*

Leaf spot *Myrothecium roridum*

Leaf spot *Leptosphaerulina trifolii*

Podrot and collar rot *Fusarium semitectum*

Thielaviopsis root rot *Thielaviopsis basicola*

Rust *Uromyces phaseoli, U. appendiculatus*

Sclerotium blight: *Sclerotinia sclerotiorum*

Dry root rot *Macrophomina phaseoli*

Anthracnose *Colletotrichum dermatum*

Anthracnose and pod blight *Colletotrichum truncatum*

Leaf spot *Mycosphaerella phaseolorum*

Leaf blight *Phytophthora vignae*

Leaf spot *Alternaria alternata*

Leaf spot *Ascochyta phaseolorum*

Leaf spot *Cercospora cruenta*

Leaf spot *Cercospora canescens*

Leaf spot *Cladosporium vignae*

Leaf spot *Corynespora cassiicola*

Root rot *Sclerotium rolfsii*

Root rot *Pellicularia filamentosa*

Rust *Uromyces phaseoli, U.appendiculatus*

Verticillium wilt *Vertcillium dahlia*

Stem canker and web blight *Rhizoctonia solani*

Bacterial diseases

Bacterial blight *Xanthomonas vignicola*

Wild fire *Pseudomonas tabaci*

Viral Diseases

Mosaic: Tobacco Streak Virus (TSV) Cowpea mosaic virus (CpMV)

LENTIL (*Lens esculenta*)

Fungal diseases

Wilt

Symptoms: The growth of the plant is checked.The roots of infected plants look light brown in colour and do not develop properly. The leaves start yellowing then the plants show wilting. Ultimately the infected plants die. Wilting is more severe in early sown crops

Causal organism: *Fusarium oxysporum* f.sp. *lentis*

Management

i) Field sanitation

ii) Crop rotation with non-host crops for minimum three years.

iii) Growing of moderately resistant varieties like Pant-4-406

iv) Use of certified seeds.

Rust

Symptoms: Brown pustules appear on leaves and stems.Then the pustules become black. The affected leaves dry up and prematurely fall.

Causal organism: *Uromyces fabae*

Management:

i) Field sanitation

ii) Growing of resistant varieties like Pant-L-406,L-9-12

iii) Use of certified seeds.

iv) Seed treatment with Vitavax Power @3g/kg of seeds.

v) Spraying of Mancozeb 50% WS (3g/litre) after appearance of the disease in the field.

Powdery mildew

Symptoms

Small whitish powdery patches appear on the surface of the leaves. Gradually these patches cover the entire leaves, young twigs and pods. Ultimately the entire plant dries.

Causal organism: *Erysiphe polygoni*

Management

i) Field sanitation

ii) Growing of resistant varieties

iii) Use of certified seeds.

iv) Seed treatment with Carboxin 37.5% WP +Thiram 37.5% WP @3g/kg of seeds

v) Spraying of Sulfex /Karathane(2g/litre) after appearance of the disease in the field.

Downy Mildew

Symptoms: On the upper surface of the leaves light greenish to yellow spots appear. The corresponding lower surface shows cottony growth of the fungus. The leaves become chlorotic.The affected plants become stunted

Causal organism: *Peronospora lentis*

Management

i) Field sanitation

ii) Growing of moderately resistant varieties like Pant-4-406

iii)Use of certified seeds.

iv) Spraying of crop with Carbendazim (12%WP)+ Mancozeb (63%W. P) @0.3%. If necessary spraying is recommended for twice /thrice at 7 days interval

Minor diseases

Fungal diseases

Seedling blight: *Botrytis cinerea*

Leaf blight: *Alternaria tenuis*

Root rot: *Pellicularia rolfsii*

Root rot: *Rhizoctonia solani*

Sclerotinia blight: *Sclerotinia sclerotiorum*

Viral diseases

Mosaic: Alfa alfa mosaic virus (AMV)

Bean yellow mosaic virus (BYMV)

Cucumber mosaic virus (CMV)

Pea leaf roll virus (PLRV)

4

Diseases of Oil Seeds

GROUNDNUT (*Arachis hypogea)*

Fungal diseases

1) Collar rot

Symptoms: The hypocotyls of germinating seeds are infected. The germinating seeds become covered with sooty black masses of spores. The infected tissues become dark brown and shredded. Necrosis and shredding of tissues may extend into branches.

The organism is both soil borne and seed borne.

Causal organism: *Aspergillus niger*

Management

i) Seed treatment with Carbendazim 50% WP @ 2-2. 5g/kg or Carboxin 37.5% WP +Thiram 37.5% WP @ 3gm/kg of seed.

ii) Growing of resistant varieties like Smruti

2) Afla rot

Symptoms: The cotyledons of germinating seeds are attacked and become necrotic.The leaflets show reduced in size and show general chlorosis and may be covered with yellowish green growth of the spores.

Causal organism: *Aspergillus flavus*

Management

i) Seed treatment with Carbendazim 50% WP @ 2-2. 5g/kg or Carboxin 37.5% WP +Thiram 37.5% WP @ 3gm/kg of seed.

ii) Growing of regional resistant varieties like Smruti

3) Charcoal Rot

Symptoms: This disease is also known as dry rot or dry root rot. This disease can occur at any time from the seedling stage to maturity. The characteristic symptom of the disease appears on the stem just above the soil level as reddish brown water-soaked necrotic lesions. The lesions spread upwards on the stem and downwards into the roots cause death of the plants. The dead tissues are covered with abundant small dark sclerotia and in some cases dark pycnidia. In advanced stages the pods become water-soaked and rotten. The kernels turn black with abundant sclerotia.

Causal organism: *Macrophomina phaseolina*

Management

i) Seed treatment with Carbendazim 50% WP @2-2. 5g/kg or Carboxin 37.5% WP +Thiram 37.5% WP @ 3gm/kg of seed.

ii) Growing of resistant varieties like TG-51

4) Diplodia Collar Rot

Symptoms: In infected plants the stems near the soil level turn grey to black. In the similar way tap roots are infected and the colour change into grey and finally become black. Individual branch may be infected. Severe infection may cause death of that branch. On necrotic tissues large number of pycnidia, i.e. are developed

Causal organism: *Diplodia gossypina*

Management

i) Clean cultivation.

ii) Growing of resistant varieties.

iii) Crop rotation with non-host crops. Soybean and cotton should not be included in crop roation

5) Rust

Symptoms: Rustic pustules develop on the lower surface of the leaves. The pustules are yellow to orange in colour about 1mm in diameter in the early stages and become brown and rough appearance after rupturing. Severely infected leaflets become necrotic, dry out and fall off prematurely.

Causal organism: *Puccinia arachidis*

Management

i) Spraying of Chlorothalonil 75% WP @ 2g/ litre/Mancozeb75% WP (0.25%) thrice in a season after 40 days of planting.

6) Stem rot

Symptoms: There is sudden wilting of plant even under humid and water saturated conditions. The leaves become chlorotic turn brown and desiccate. Abundant white mycelial growth develop on the stem near soil level

Causal organism: *Sclerotium rolfsii*

Management

i) Seed treatment with Carbendazim 2-2. 5g/kg or Carboxin 37.5% WP +Thiram 37.5% WP @3gm/kg of seed.

ii) Growing of resistant varieties like TG-51

iii) Crop rotation with wheat and maize may reduce the disease incidence

7) Leaf spots

Symptoms: On the leaves, oval, elongated spots appear in early leaf spot disease. They may coalesce to form irregular lesions. These lesions are dark brown to black colour on the upper surface while light brown colour on the lower surface. There may be presence of yellow halo around the lesions on the upper surface. In severe cases petioles and stems are affected. In late leaf spot the dark brown to black coloured lesions are produced on stems petioles and leaf lets. These lesions are 1-6 mm in diameter. These lesions may coalesce to form irregular necrotic areas.

Causal organism

i) *Cercospora arachidicola*

ii) *C. personata / Phaeoisariopsis personata*

Management

i) Clean cultivation

ii) Destruction of volunteer plants

iii) Use of certified seeds

iv) Spraying of with Carbendazim 12%+Mancozeb 63% WP (0.2%) thrice in a season after 40 days of planting at 10-15 days interval

Viral diseases

1) Bud Necrosis/ Bud Blight

Symptoms: The affected plants show stunted growth. There is necrosis is of terminal bud and proliferation of auxillary shoots with deformed leaves. Yield is reduced.In some cases plants produce small discoloured and deformed kernels

Causal organism: Groundnut bud necrosis virus (GBNV)

Vector: *Thrips palmi*

Management

i) Clean cultivation

ii) Destruction of volunteer plants

iii) Use of certified seeds

iv) Rouging of diseased plants. High planting density should be maintained to limit the infestation.

v) Spraying of Dimethoate 30% EC (400 ml/ha) twice or thrice at 7 days interval for controlling of vectors should be practised.

2) Mottling

Symptoms: The young leaves of affected plants show mild mottling and vein clearing while older leaves show upward curling and interveinal depressions.

Causal organism: Groundnut mottle virus

Vector: Aphid (*Myzus persicae, Aphis craccivora)*

Management

i) Use of certified seeds

ii) Rouging of diseased plants

iii) Spraying of crop with Imidacloprid 17.8% SL@1.0 ml/3 litre or Thiomethoxam 25% WG@1.0g/5 litre

3) Rosette

Symptoms: The affected plants show bushy appearance and remain stunted. Leaves may show mosaic and mottling symptoms and there is remarkale reduction in size of the leaves.

Causal organism: Groundnut rosette virus (GRV)
Groundnut rosette assistor virus (GRAV)

Vector: Aphid (*Aphis craccivora,A.gossypii)*

Management

i) Use of certified seeds

ii) Rouging of diseased plants

iii) Spraying of crop with Imidacloprid 17.8% SL@1.0 ml/3 litre or Thiomethoxam 25% WG@1.0g/5 litre

4) Clump

Symptoms: The young leaves show mosaic mottled leaves. Plant height reduction and roots become dark coloured. The virus has wide host range and it was reported to be transmitted by nematodes.

Causal organism: Indian Peanut Clump Virus (IPCV)

Management

i) Use of certified seeds

ii) One week before planting Nemagon (0.2%) should be applied.

5) Yellow spot

Symptoms: It is a minor disease.The height of the plants reduced. Yellow spots with light brown margin appear on leaves. The leaf and leaf lets show reduction in size.

Causal organism: Peanut yellow spot virus (PYSV)

Vector: Thrips *Frankliniella occidentalis F. fusca, scirtothrips dorsalis*

Management

Use of certified seeds

Minor diseases

Fungal diseases

Damping off *Fusarium oxysporum*

Seedling blight: *Pythium debaryanum*

Seed and seedling rot *Rhizopus arrhizus*

Root rot and cotyledon rot: *Pythium myriotylum*

Anthracnose:

i) *Colletotrichum capsici*

ii) *C. dematium*

iii) *C.gloeosporiodes*

Botrytis blight *Botrytis cinerea*

Leaf spot *Alternaria alternata*

Leaf spot *A. arachidis*

Leaf spot *Leptosphaerulina trifolii*

Leaf spot *Pestalotiopsis arachidis*

Leaf spot *Phoma microspora*

Leaf spot *Phyllosticta arachis hypogea*

Leaf spot *Curvularia lunata*

Diplodia collar rot *Diplodia gossypina*

Collar rot *Diplodia arachidis*

Stem rot *Botryobasidum rolfsii*

Verticillium wilt and rot: *Verticillium albo-atrum*

SESAMUM (*Sesamum indicum*)

Fungal diseases

1) Sclerotium rot

Symptoms: A whitish mycelial mat is formed around the collar region of the plant due to infection at the collar region of the plant and later extend to the roots causing defoliation and death of the plants.

Causal organism: *Sclerotium rolfsii*

Management

i) Seed treatment with Carbendazim 50% WP @ 2-2. 5g/kg

ii) Basal drenching with Carbendazim 50% WP @ 1.5gm/litre

iii) Removal of crop debris after harvesting of the crop

iv) Crop rotation

2) Alternaria leaf spot

Symptoms: The disease appears as irregular brown spots with concentric rings on leaves and in severe cases elongated lesions appears on stems and capsules. The plants are killed due to severe defoliation and stem infection.

Causal organism: *Alternaria sesami*

Management

i) Seed treatment with Carbendazim 50%WP 2-2. 5g/kg or Carboxin 37.5% WP +Thiram 37.5% WP @3gm/kg of seed.

ii) Spraying of crop with Mancozeb 75% WP (0.2%) twice/thrice at 7 days interval after appearance of the disease in field.

3. Cercospora leaf spot

Symptoms: The disease symptoms appear at the time of flowering. There are light brown spots on both the surface of the leaf. These spots coalesce with each other which increase in size to form bigger patches on the leaves. Severe leaf infection cause defoliation of the plants.

Causal organism: *Cercospora sesami*

Management

i) Seed treatment with Carbendazim 50%WP @ 2-2. 5g/kg or Carboxin 37.5% WP +Thiram 37.5% WP @3gm/kg of seed.

ii) Spraying of crop with Mancozeb 75%WP (90.2%) twice or thrice at 7 days interval after appearance of the disease in field.

4. Charcoal rot

Symptoms: The roots and collar region of the stem show brown colouration which later on extends upward. Gradually the whole plant becomes brown coloured. Minute pycnidial bodies are seen all over the stem. Severely infested plants fail to produce seeds in rare cases small shrivelled seeds are produced.

Causal organism: *Macrophomina phaseolina*

Management

i) Seed treatment with Carbendazim 50%WP @ 2-2. 5g/kg or Carboxin 37.5% WP +Thiram 37.5% WP @ 3gm/kg of seed.

ii) Growing of resistant varieties, i.e.Gwalior-5, ST-58 and TNAU-2

5. Stem Blight

Symptoms: Small water soaked irregular spots appear on the leaves which later extend,enlarge and cover the entire leaf area. The infected leaves become flaccid, droop and finally wither. Blackening of the stem at soil level is the typical symptom of the disease. The infected plants fail to produce flower and die prematurely.

Causal organism: *Phytophthora parasitica* var. *sesami*

Management

i) Seed treatment with Carbendazim 50% WP @ 2-2. 5g/kg or Vitavax Power @ 3gm/kg of seed.

ii) Growing of resistant varieties, i.e.Vinayak

iii) Spraying of Metalaxyl M 4% WP + Mancozeb 64% WP(0.3%) twice or thrice at 7-10 days interval after appearance of the disease in the crop.

6. Powdery Mildew

Symptoms: The affected leaves are covered with whitish fungal growth and fall off prematurely.

Causal organism: *Erysiphe polygoni*

Management

i) Clean cultivation

ii) Growing of resistant varieties

iii) Spraying of crop with Sulphur 80% WP @ 0.4% or dusting with the Sulphur 85% DP @ 20kg/ha at 45 th and 65 th days of the crop sowing.

7. Wilt

Symptoms: The characteristic symptom is the wilting and drying of the plants. Yellowing, withering and drying of leaves occur followed by drying of the entire plant. There is discolouration of stem and disintegration of cortical tissues.

Causal organism: *Fusarium oxysporum* f. sp. *sesami*

Management

i) Clean cultivation

ii) Crop rotation for two years.

iii) Growing of resistant varieties likeAmrit

Bacterial diseases

1. Bacterial leaf spot

Symptoms: Small water soaked spots develop on the cotyledons of seedlings. In older seedlings dark brown to black colour spots of round to irregular shape appear on the true leaves. These spots coalesce to form bigger patches and cause defoliation of leaves.

Causal organism: *Xanthomonas sesami*

Management

i) Seed treatment with Streptocycline 250ppm. (soaking of seeds for 30 minutes before sowing)

ii) Growing of resistant varieties i.e. Nirmala

iii) Spraying of Streptocycline 250ppm or Copper oxychloride @ 0.2% twice or thrice at 7 days interval after appearance of the disease in the field.

Viral diseases

1. Leaf curl

Symptoms: The typical symptom is the brittleness, reduction in size and downward curling of the infected leaves.

Causal organism

Necotiana virus-10

Vector: Aphid (*Myzus persicae, Aphis craccivora*)

Management

i) Removal of crop debris after harvest of the crop

ii) Use of certified seeds.

iii) Growing of resistant varieties like Prachi,Nirmala.

iii) Spraying of Imidacloprid 17.8% SL@1.0 ml/3 litre or Thiomethoxam 25% WG@1.0g/5 litre of water will prevent the disease spread.

Phytoplasma diseases

1) Phyllody

Symptoms: The affected plant parts are malformed. The inflorescence are transformed to leafy structure.

Causal organism

Mycoplasma like organism

Vector: Jassid (*Orosius albicinctus*)

Management

i) Removal of crop debris after harvest of the crop

ii) Use of certified seeds.

iii) Growing of resistant varieties like Prachi, Nirmala.

iii) Spraying of Imidacloprid 17.8% SL@1.0 ml/3 litre or Thiomethoxam 25% WG@1.0g/5 litre of water will prevent the disease spread.

Minor diseases

Fungal diseases

Drechslera blight *Helminthosporium sesami*

Leaf spot: *Cercospora sesamicola*

Leaf spot: *Phoma exigua*

Stem rot: *Phytophthora parasitica* var *sesami*

Pre-emergence seedling blight: *Pythium* sp

Leaf spot: *Phyllosticta sesami*

Leaf spot: *Curvularia lunata*

Leaf spot: *Myrothecium roridum*

Leaf spot: *Pseudocercosporella sesami*

Leaf and stem gall: *Synchytrium sesame*

*L*eaf and stem gall: *Synchytrium sesamicola*

Powdery mildew: *Leveillula taurica*

Wilt: *Fusarium vasinfectum*

Verticillium wilt: *Verticillium albo-atrum*

Bacterial diseases:

Bacterial blight *Pseudomonas sesami*

Viral diseases

Mosaic

SUNFLOWER (*Helianthus annus)*

Fungal diseases

1) Leaf spot

Symptoms: The symptom of the disease first appears on the leaves as brown coloured spots. These spots increases in size become dark brown lesions sourrounded by yellow halo. In severe infection leaf blightening occurs and brown coloured spots appear on stem, petals and sepals.

Causal organism: *Alternaria helianthi*

Management

i) Seed treatment with Carbendazim 50%WP @ 2-2. 5g/kg or Carboxin 37.5% WP +Thiram 37.5% WP @ 3gm/kg of seed.

ii) Spraying of crop with Mancozeb 50%WS (0.2%) thrice in a season after 30, 40 and 50 days after planting.

iii) Growing of moderately resistant varieties, i.e. KBSH-1, KBSH-44, KBSH-41

2) Sclerotium wilt

Symptoms: The disease affects the plants at any stage of growth. The affected plants show a sickly appearance. A whitish fungal growth with sclerotia(mustard seed like structure) on the stem of the affected plant near the soil surface can be seen.

Causal organism: *Sclerotium rolfsii*

Management

i) Use of certified seeds

ii) Field sanitation. The fields should be free from weed.

iii) Removal and destruction of diseased crop debris

3) Sclerotinia rot

Symptoms: Rotting of head and stem can be observed.

Causal organism: *Sclerotinia sclerotiorum*

Management

i) Seeds from healthy crop can be used.

ii) Certified seeds should be used for planting.

iii) Clean cultivation

iv) Field sanitation. The fields should be free from weed.

v) Spraying of crop with Mancozeb 75% WP (0.2%) at the time of head initiation has been effective in controlling the disease.

4) Powdery mildew

Symptoms: White powdery growth appear on the surface of leaves, young shoots and flowers. In severe the leaves and other infected parts dry up.

Causal organism: *Erysiphe cichoracearum*

Management

i) Removal of crop debris after harvest of the crop

ii) Use of certified seed

iii) Spraying of Sulphur 80% WP @ 0.4 % twice or thrice at 7 days interval after appearance of the disease.

5) Head Rot

Symptoms: Irregular shaped several water-soaked spots develop on the surface of the head adjacent to the stalk and gradually spread to the head. The under surface of the head become very sof pulpy and discoloured . Insects make small holes on head. The pathogen enters through holes and attack the head and seeds. Finally the seeds are converted into black powdery mass.

Causal organism: *Rhizopus* sp

Management

i) Clean cultivation

ii) Field sanitation. The fields should be free from weed.

iii) Spraying of crop with Mancozeb 75% WP @ (0.2%) at the time of head initiation has been found effective in controlling the disease.

6) Rust

Symptoms: Small reddish-brown pustules develop on the lower surface of the leaves. The pustules may cover the entire leaf. All the leaves, stem, branches and green parts of flowers are also affected. The leaves turn yellow.

Causal organism: *Puccinia helianthi*

Management

i) Spraying of Chlorothalonil 75%WP @ 0.2% or Difenoconazole (0.05%) thrice in a season after 40 days of planting.

ii) Growing of moderately resistant varieties, i.e., KBSH-1, KBSH-44, KBSH-41

7) Charcoal Rot

Symptoms: Black ashy growth appear on the stem. The affected plants mature early and produced under sized heads

Causal organism: *Macrophomina phaseoli*

Management

i) Removal of crop debris after harvest of the crop

ii) Use of certified seed

Minor diseases

Fungal diseases

Downy mildew *Plasmopara halstedii*

Leaf spot and blight *Alternaria alternata*

Leaf and stem spot *A. zinnae*

Leaf spot *Cladosporium cladosporiodes*

Leaf spot *Helminthosporium helianth*i

Leaf spot *Septoria helianthi*

Leaf blight *Botryodiplodia theobromae*

Powdery mildew *Spacelotheca fuliginea*

Root rot *Rhizoctonia bataticola*

Vericillium wilt *Verticillium alboatrum*

White blister *Albugo* sp

Viral diseases

Mosaic Sunflower mosaic virus

Mosaic Poppy mosaic virus

MUSTARD AND RAPESEED (*Brassica campestris, Brassica juncea*)

1) Alternaria blight

Symptoms: Production of small brown necrotic spots with yellow halo on the leaves. These spots increase in size and become light brown to dark brown or black coloured. In these spots concentric rings are present. In severe infection blightening occurs and stems and pods are infected. In severe infection seeds are shrivelled and become undersized.

Causal organism: *Alternaria brassicae*

Management

i) Collection and destruction of diseased plant parts.

ii) Healthy seeds used for planting.

iii) Seed treatment with Carbendazim 50%WP @2-2. 5g/kg or Carboxin 37.5% WP +Thiram 37.5% WP @3gm/kg of seed.

iv) Spraying of Mancozeb (0.3%) twice or thrice at 7 days interval after appearance of the disease.

v) Growing of resistant varieties like Parbati and Anuradha

2) Stem rot or Stem blight

Symptoms: Water soaked lesions appear on stem which covered by white fungal growth on which black sclerotia appear. The seeds may be shrivelled.

Causal organism: *Sclerotinia sclerotiorum*

Management

i) Seeds from healthy crop can be used.

ii) Certified seeds should be used for planting.

iii) Clean cultivation.

iv) Field sanitation. The fields should be free from weed.

3) Downy Mildew

Symptoms: All aerial parts are infected. In severe infection inflorescence is affected. Yellow patches develop on leaves. The corresponding lower suface shows whitish mycelial growth. In severe infection the inflorescence is malformed, twisted and there is no pod formation. Some times the stems swell and may be smaller or several inches longer.

Causal organism: *Peronospora brassicae*

Management

i) Use of healthy seeds for sowing.

ii) Seed treatment with Carbendazim 50% WP @2-2. 5g/kg or Carboxin 37.5% WP +Thiram 37.5% WP @3gm/kg of seeds.

iii) Growing of resistant varieties like Parbati and Anuradha.

iv) Use of certified seeds

v) Seed treatment with Carboxin 37.5% WP +Thiram 37.5% WP @ 3gm/kg

vi) Spraying of (0.1%) twice or thrice at 7 days interval after appearance of the disease.

3) White rust/White blister

Symptoms: White or creamy yellow pustules appear scatteredly on the under surface of the leaves. These blisters burst and liberate white powdery mass of spones. In severe infection stems and flowers are also affected. The flowers get malformed and become sterile.

Causal organism: *Albugo candida*

Management

i) Removal of crop debris after harvest of the crop.

ii) Crop rotation with non-cruciferous crops for two years

iii) Use of certified seeds

iv) Spraying of Mancozeb 50% WS (0.3%) twice or thrice at 7 days interval as soon as the symptoms are noticed in the field.

4) Powdery mildew

Symptoms: Appearance of white floury/powdery patches on both the sides of leaves as well as on the green parts of the plants. These floury/powdery patches consist of mycelium and conidial stage of the fungus. In advanced stage the entire surface may be covered with white powdery like mass. Later on, black dots may appear late in the season.The disease affects pod formation.

Causal organism: *Erysiphe polygoni*

Management

i) Removal of crop debris after harvest of the crop

ii) Use of certified seeds

iii) Spraying of Sulphur 80% WP @0.2% twice or thrice at 7 days interval after appearance of the disease.

Phytoplasma diseases

1) Phyllody

Symptoms: The affected plant parts are malformed. Numerous shoots arise and the plants become broom shaped. The inflorescences are transformed to leafy structures.

Causal organism: Mycoplasma like organism

Management

i) Removal of crop debris after harvest of the crop

ii) Use of certified seeds

iii) Avoid early sowing of crop

iv) Rogue out the affected plants early.

Minor diseases:

Fungal diseases

Damping off *Pythium* sp, *Fusarium* sp, *Phytophthora* sp

Dark leaf spot *Alternaria brassicicola*

Leaf spot *Cercospora brassicicola*

Black leg *Phoma lingam*

Root gall smut *Urocystis brassicae*

Stem and root rot *Macrophomina phaseoli*

Rizoctonia blight *Rhizoctonia solani*

Fusarial wilt *Fusarium oxysporum* f.sp. *coglutinans*

Bacterial diseases

Black rot *Xanthomonas campestris*

Viral diseases

Mosaic: Watermelon mosaic virus (WMV)

NIGER (*Guizotia abyssinica*)

1) Alternaria leaf spot

Symptoms: Small brown circular spots are are formed on the leaves. These spots gradually increase in size and cover the entire leaf. The infected leaves turn yellow and drop prematurely.

Causal organism: *Alternaria* sp

Management

i) Removal of crop debris after harvest of the crop

ii) Use of certified seeds

iii) Spraying of Mancozeb 50%WS (0.3%) twice or thrice at 7 days interval as soon as the symptoms are noticed in the field.

iv) Growing of resistant varieties like Deomali.

2) Root rot

Symptoms: Roots turn black and rot. Infected plants show wilting.

Causal organism: *Rhizoctonia bataticola*

Management

i) Growing of resistant varieties like Deomali

ii) Seed treatment with Carbendazim 50% WP @ 2-2. 5g/kg or Carboxin 37.5% WP +Thiram 37.5% WP @3gm/kg of seed.

iii) Soil treatment

3) Powdery mildew

Symptoms: Powdery or floury patches appear on leaves, young shoots and flowers. Ultimately the leaves dry up.

Causal organism: *Oidium* sp

Management

i) Removal of crop debris after harvest of the crop

ii) Use of certified seeds

iii) Spraying of Carbendazim 12% +Mancozeb 63% WP (0.2%) twice or thrice at 7 days interval after appearance of the disease.

Minor diseases

Fungal diseases

Leaf spot *Alternaria alternata*

Leaf spot *Cercospora guizoticola*

Leaf spot *Ozonium taxanum* var.*parasiticum*

Rust *Puccinia guizotiae*

Bacterial diseases

Bacterial leaf spot *Xanthomonas campestris* pv.*guizotiae*

LINSEED (*Linum usitatisimum*)

Fungal diseases

1) Rust

Symptoms: Minute pustules of bright orange coloured are produced on leaves and stems. As the pustules advance the pustules turn black.

Causal organism: *Melamposora lini*

Management

i) Seed treatment with Carbendazim 50%WP @2-2. 5g/kg or Carboxin 37.5% WP +Thiram 37.5% WP @ 3gm/kg of seed.

ii) Growing of resistant varieties like Hira, Neelum Mukta, Mayurbha LC -0185

iii) Spraying of Chlorothalonil 75% WP 0.2% or Difenaconazole (0.05%) thrice in a season after 40 days of planting.

2) Blight

Symptoms: Brown coloured spots appear on the leaves. These spots increases in size. Several spots appear on the leaves and cover the entire leaves resulting blightening of leaves.

Causal organism: *Alternaria lini*

Management

i) Seed treatment with Carbendazim 50%WP @2-2. 5g/kg or Carboxin 37.5% WP +Thiram 37.5% WP @ 3gm/kg of seed.

ii) Growing of resistant varieties.

iii) Spraying of crop with Mancozeb 50%WS (0.2%) thrice in a season 30, 40 and 50 days after planting.

Minor diseases

Fungal diseases

Root rot	*Pythium aphanidermatum*
Seed rot	*Rhizoctonia* sp
Seed rot	*Stemphylium botryosum*
Leaf spot	*Alternaria alternata*
Leaf spot	*Cercospora linicola*
Leaf and stem spot	*Pleospora* sp
Powdery mildew	*Oidiopsis lini*
Stem and root rot	*Pellicularia rolfsii*
Stem and root rot	*Sclerotinia sclerotiorum*

Bacterial diseases

Bacterial leaf spot: *Xanthomonas campestris* pv. *lini*

CASTOR (*Ricinus communis*)

1) Seedling blight

Symptoms: Roundish patch of dull green colour first develops on both the surfaces of cotyledonary leaves. Later the spots turn yellow or brown. Under humid condition under surface of the leaves become covered with whitish fungal growth. In severe infection growing point of leaves shows rotting.

Causal organism: *Phytophthora parasitica*

Management

i) Use of certified seeds

ii) Seed treatment

iii) Growing of resistant varieties like TMV-9,HC-6 and PC-1

2) Alternaria Blight

Symptoms: Small brown necrotic spots appear on all the above ground parts of the plant. These spots increase in size and become light brown to dark brown or black cause blightening of leaves. In necrotic lesions concentric rings are present.

Causal organism: *Alternaria ricini*

Management

i) Use of certified seeds

ii) Seed treatment with Carbendazim 50% WP (0.2%)

iii) Spraying of Mancozeb 50%WS (0.3%) twice or thrice at 7 days interval as soon as the symptoms are noticed in the field.

3) Rust

Symptoms: Small reddish brown pustules develop on the lower surface of the leaves. The pustules may cover the entire leaf. All the leaves, stem, branches and green parts of flowers are also affected. The leaves turn yellow.

Causal organism: *Melampsora ricini*

Management

Spraying of Chlorothalonil (0.03%) or Mancozeb (0.2%) thrice in a season after 40 days of planting.

4) Powdery mildew

Symptoms: Powdery or floury patches appear on leaves, young shoots and flowers. Ultimately the leaves dry up.

Causal organism: *Leveillula taurica*

Management

i) Removal of crop debris after harvest of the crop

ii) Use of certified seeds

iii) Spraying of Karathane (0.2%) twice or thrice at 7 days interval after appearance of the disease.

5) Twig blight

Symptoms: Oval to spherical spots with dark brown margin appear on stem, branches and petioles. The centre of the spots become necrotic. Later on these spots enlarge and the centre of the spots become whitish on which black fruiting bodies appear.

Causal organism: *Glomerella cingulata*

Management

i) Use of certified seeds

ii) Seed treatment with Carbendazim 50% WP (0.2%)

iii) Spraying of Mancozeb 50% WS @ (0.3%) twice or thrice at 7 days interval as soon as the symptoms are noticed in the field.

6) Leaf spot

Symptoms: Minute black or brown spots with pale green margin appear on the leaves. Later these spots become brown and then greyish white with deep brown margin. On the centre of the spots black tiny dots present.

Causal organism: *Cercospora ricinella*

Management

i) Use of certified seeds

ii) Seed treatment with Carbendazim 50%WP @ (0.2%)

iii) Spraying of Mancozeb 50% WS (0.3%) twice or thrice at 7 days interval as soon as the symptoms are noticed in the field.

7) Pod rot

Symptoms: The infection is cofined to castor pods. The capsules cracks at point of infection. On the seeds water soaked areas develop which subsequently rot

Causal Organism: *Cladosporium oxysporum*

Management

i) Use of certified seeds

ii) Seed treatment with Carbendazim 50%WP (0.2%)

iii) Spraying of Mancozeb (0.3%) twice or thrice at 7 days interval as soon as the symptoms are noticed in the field.

8) Bacterial leaf spot

Symptoms: Small brown water soaked spots appear on leaves which become angular and dark brown in colour. In severe infection leaf blightening occurs.

Causal organism: *Xanthomonas ricinicola*

Management

i) Removal of crop debris after harvest of the crop

ii) Use of certified seeds

Minor diseases

Fungal diseases

Damping off: *Pythium debaryanum*

Leaf spot: *Alternaria tenuissima*

Leaf spot: *Corynespora cassicola*

Leaf spot: *Leptosphaeria ricini*

Root and Stem rot: *Macrophomina phaseolina*

Fruit rot: *Cladosporium oxysporum*

5

Diseases of Fiber Crops

COTTON (*Gossypium* spp)

Fungal diseases

1) Damping off

Symptoms: The seeds may rot. Initially the roots and hypocotyls become water soaked and pale green in colour. The hypocotyls become constricted above the soil level and seedlings fall down.

Causal organism: *Pythium aphanidermatum*

Management

i) Seed treatment

ii) Growing of resistant varieties.

2) Root rot

Symptoms: Roots of infected plants rot and show shredding of the bark. The affected plants can be easily pulled out.

Causal organism: *Rhizoctonia solani*

Management

i) Seed treatment

ii) Growing of resistant varieties.

3) Fusarium wilt

Symptoms: The plants show gradual wilting. Yellowing and drooping of leaves entire plant show wilting.

Causal organism: *Fusarium oxysporum* f. sp. *vasinfectum*

Management

i) Crop rotation (3 to 4 years)

ii) Seed treatment.

iii) Growing of resistant varieties.

4) Leaf spots

i) Alternaia leaf spot

Symptoms: Typical brown mostly circular lesions with concentric rings and reddish border appear on affected bolls,bracts and leaves particularly very young leaves. These lesions may coalesce and form larger lesions. Ultimately extensive defoliation occurs in this disease.

Causal organism: *Alternaria macrospora*

ii) Cercospora leaf spot

Symptoms: The leaf spots are small,round to irregular with white centres surrounded by purple border. As the lesion enlarges the centre may drop out giving a shot hole effect. The affected leaves turn brown and dry up resulting partial or complete defoliation

Causal organism: *Cercospora gossypina*

iii) Helminthosporium leaf spot

Symptoms: Numerous light brown circular spots appear on the leaves, flower bracts and bolls. Severe leaf infection cause blightening of leaves.

Causal organism: *Helminthosporium gossypii*

Management

i) Removal of crop debris after harvest of the crop

ii) Use of certified seeds

iii) Spraying of Carbendazim 12% +Mancozeb 63% WP @ (0.2%)twice or thrice at 7 days interval after appearance of the disease.

5) Powdery Mildew

Symptoms: White powdery patches develop on surface of leaves. The young shoots may be affected. The infected leaves may dry up.

Causal organism: *Leveillula taurica*

Management

i) Spray the crop with Wettable Sulphur 80%WG@ @2.5g/litre

6) Grey mildew

Symptoms: Yellowish green,irregular,angular lesions appear which are delimited by veins. On the lower surface of these lesions there is profuse fungal growth giving whitish appearance. Severe infection may result in defoliation,premature opening of bolls and deterioration of fibres.

Causal organism: *Ramularia aerola*

Management

i) Removal of crop debris after harvest of the crop

ii) Use of certified seeds

iii) Spraying of Copper oxychloride 50%WP @3g/litre twice or thrice at 7 days interval after appearance of the disease.

7) Rust

Symptoms: Numerous small yellowish brown rust pustules (pycnia) are formed on the leaves. The uredia spores are developed which are light cinnamon-brown while telia are dark brown. In severe infestations the plants may be defoliated.

Causal organism: *Puccinia stakmanii*

Management

i) Removal of crop debris after harvest of the crop

ii) Use of certified seeds

iii) Spraying of Carboxin or Oxycarboxin (0.2%) or Propiconazole 50%WP @1.5 /litre twice or thrice at 7 days interval after appearance of the disease.

8) Bacterial Blight (Black arm or Angular leaf spot)

Symptoms: The disease attack all above ground parts of the plant at all stages of plant growth. The earliest symptom of the disease is seen as round water-soaked green spots of different sizes on the under surface of cotyledons. The disease spreads to newly emerged leaves formed. Ultimately the seedlings may collapse and die.

The angular and water-soaked brown to black lesions delimited by small veinlets are observed on the leaves. Several lesions coalesce and cause death and shedding of leaves.

The lesions on the stems and petioles are dark brown to sooty black elongated and sunken. The affected stems show cracks and gummosis and are easily broken by wind or there may be girdling or death of the affected organs.

Lesions on bolls cause premature dropping.

Causal organism: *Xanthomonas campestris* pv *malvaearum*

Management

i) Use of certified seeds

ii) Growing of resistant varieties.

iii) Soak the seeds in 10 litres of water containing 0.5gm of Streptocycline and 10gm of Carbendazim for 6 to 8 hours. One gram of succinic acid is added in this solution to promote good crop establishment.

iv) Spray the crop with Copper oxychloride 50% WP @ 2.5-3gm/litre of water after 45 days of sowing twice or thrice at 15 days interval.

v) Removal of crop debris after harvest of the crop

Viral Diseases

i) Stenosis or Small Leaf

Symptoms: The sizes of the leaves are drastically reduced and shape disorted. The size of flowers reduced. Ovaries are damaged. Hence bolls are not formed. Some times only one or few branches are affected.

Causal organism: Cotton small leaf virus

Management

i) Use of certified seeds

ii) Growing of resistant varieties.

Nematode diseases

i) Reinform Nematode

Symptoms: Roots of the plants in seedling stage are damaged. The plants show stunted growth. The lateral roots show stubby structure.

Causal organism: *Rotylenchulus reniformis*

Management

i) Use of certified seeds

ii) Growing of resistant varieties.

iii) Crop rotation with rice, jowar, sugarcane, onion,mustard,chilli and turnip

iv) Application of DBCP@10kg/ha

Non-parasitic diseases

1) Red leaf disease

Symptoms: The plants are attacked in all stages. The leaves turn yellow or red and roll downwards. In case of severe infection the leaf lamina becomes red except veins which are remain green ultimately shed. This is often followed by shedding of buds, flowers and young bolls.

Cause: Physiological

Management

i) Use of certified seeds

ii) Growing of resistant varieties.

iii) Use of proper dose of NPK fertilizers

2) Tirak

Symptoms: The leaves show yellowing, droop and untimely shade. The bolls show cracking and do not open properly. The seeds can not mature properly and the lint quality become poor

Cause: Physiological

Management

i) Use of certified seeds

ii) Growing of resistant varieties.

iii) Use of balance dose of NPK in light sandy and alkaline soil extra irrigation is recommended.

Minor diseases

Fungal diseases

Damping off and Boll rot: *Phytophthora parasitica*

Boll rot: *Rhizopus stolonifer*

Alternaria leaf spot: *Alternaria gossypina*

Leaf spot: *Alternaria alternata*

Leaf spot: *Cercospora gossypii*

Leaf blight: *Phoma exigua*

Anthracnose: *Colletotrichum capsici*

Diplodia boll rot: *Diplodia gossypina*

Nigrospora lint rot: *Nigrospora oryzae*

Pestalotia leaf spot: *Pestalotia gossypii*

Verticillium wilt: *Verticillium dahliae*

Viral diseases

Mosaic Tobacco mosaic virus (TMV)

JUTE (*Corchorus capsularis* and *C. olitorius*)

1) Anthracnose of Capsularis

Symptoms: Tiny brown to black spots appear on the stem which later coalesce to form larger necrotic lesion. The necrotic spots become dark, dry and cankerous in advanced stage of the disease. The fibres are frequently shredded and the quality of fibre is poor.

Causal organism: *Colletotrichum corchorum*

Management

i) Seed treatment

ii) Spray the crop with Copper oxychloride 50% WP@ 3g/litre of water

2) Anthracnose of Olitorius

Symptoms: Many dark spots appear on the stem at about three months stage of the crop. In severe cases these spots coalesce to form larger necrotic lesion. The necrotic spots become dark, dry and cankerous in advanced stage of the disease. The quality of fibre become poor.

Causal organism: *C. gloeosporioides*

Management

i) Seed treatment

ii) Spray the crop with Copper oxychloride 50%WP @ 3g/litre of water

3) Macrophomina disease

Symptoms: The fungus cause seedling blight, root rot and stem rot in plants

i) **Seedling blight:** Seedlings develop black lesions on cotyledons which rot and wither. The seedlings die ultimately and the fields look blighted

ii) **Root rot**: The tips of affected plants droop and the colour of the stem gradually turns brown and finally the the plants look dry and wither away. The infected roots shows rotting becomes rusty brown.

iii) **Stem rot:** The black lesions appear at nodes and spread upward and downward along the stem. The black coloured minute sclerotia are developed on these lesions. The plants are usually killed.

Causal organism: *Macrophomina phaseolina*

Management

i) Use of certified seeds

ii) Growing of resistant varieties.

iii) Spraying of Copper oxychloride 50%WP @3g/litre twice or thrice at 7 days interval after appearance of the disease.

3) Powdery Mildew

Symptoms: White powdery patches develop on both the surface of leaves. In severe infection the leaves become dried.

Causal organism: *Oidium* sp

Management

i) Spray the crop with Sulphur 80% WP @ 4g/litre

4) Cercospora leaf spot

Symptoms: Small dark brown coloured spots appear on the leaves. These spots increase in size. Several spots appear on the leaves and cover the entire leaves resulting blightening of leaves. Later dark lesions develop on stem which result in discolouration of fibres.

Causal organism: *Cercospora corchori*

Management

i) Seed treatment with Thiram 75% WS/ Carbendazim 50% WP @2-2.5g/kg or Carboxin 37.5% WP +Thiram 37.5% WP @ 3gm/kg of seed.

ii) Spraying of crop with Mancozeb 50%WS (0.2%) thrice in a season at 10 days interval.

Bacterial diseases

1) Bacterial wilt

Symptoms: Suddenly wilting of plants are observed.

Causal organism: *Ralstonia solanacearum*

Management

i) Use of certified seeds

ii) Growing of resistant varieties.

iii) Seed treatment with Copper oxychloride @0.2%

iv) Application of stable bleaching powder @ 12kg/hectare

Viral diseases

1) Chlorosis or Leaf Mosaic

Symptoms: Mosaic, mottling and chlorosis of leaves occur. In severe infection plants remain stunted and ultimately killed.

Causal organism: Corchorus golden mosaic virus (CoGMV)

Management

i) Use of certified seeds

ii) Growing of resistant varieties (Olitorius varieties)

iii) Spraying the crop with Imidacloprid 17.8% SL @2. 5ml/10litre of water twice.

Minor diseases

Fungal diseases

Damping off *Pythium* sp.

Leaf spot *Cercospora capsularis*

Leaf spot *Cercospora macutensis*

Leaf spot *Phyllosticta* sp

Leaf spot *Myrothecium roridum*

Rhizoctonia leaf spot *Thanatephorus cucumeris*

Leaf and fruit spot *Alternaria citri*

Powdery mildew *Oidium* sp

Wilt *Fusarium solani*

Stem and root rot *Sclerotium rolfsii*

SUNNHEMP (*Crotolaria juncea)*

Fungal diseases

Wilt

Symptoms: The leaves become yellow and the vascular tissues of stem and roots are discoloured.The affected plants dry up.

Causal organism: *Fusarium oxysporum* var. *crotolariae*

Fusarium vasinfectum

Management

i) Use of certified seeds

ii) Growing of resistant varieties like K-12, Black, K-12 Yellow, Chindwada, Nalanda Sanai etc.

iii) Seed treatment with Thiram 75%WS @3g/kg or Carboxin 37.5% WP+Thiram 37.5% WP 23gm/kg of seed.

iv) Crop rotation with non-host crops for 3to 4 years

Rust

Symptoms: The yellow brown pustules appear on aerial parts of the plant which become dark brown in later stage. Ultimately the affected plants dry up.

Causal organism: *Uromyces* sp

Management

i) Use of certified seeds

ii) Growing of resistant varieties.

iii) Spray the crop with Mancozeb 75%WP(0.3%) or Carbendazim 50%WP (0.2%) twice or thrice at 7 days interval

Anthracnose

Symptoms: The black lesions develop on cotyledons which drop out from the seedlings. Similarly leaves are also affected.

Causal organism: *Colletotrichum curvatum*

Management

i) Use of certified seeds

ii) Seed treatment with Thiram 75%WS @3g/kg or Carboxin 37.5% WP+Thiram 37.5% WP 23gm/kg of seed.

iii) Growing of resistant varieties.

iv) Spray the crop with Mancozeb 75%WP (0.3%) or Carbendazim 50%WP (0.2%) twice or thrice at 7 days interval

Mosaic

Symptoms: The leaves show mosaic patches and become motted. The severely infected plants are stunted.

Causal organism: Southern sunnhemp mosaic virus (SHMV)

Management

i) Use of certified seeds

ii) Growing of resistant varieties.

iii) Spraying of crop with Imidacloprid (0.025%) to check the vector.

Phyllody

Symptoms: The plants become stunted. The size of the leaves is reduced.The floral parts are transformed into leafy structures.

Causal organism: Phytoplasma

Management

i) Use of certified seeds

ii) Growing of resistant varieties.

iii) Spraying of crop with Thiomethoxam @ 1g/5litre of water to check the vector.

Leaf curl

Symptoms:The leaves become curled. The plants remain dwarf.

Causal organism: Tobacco leaf curl virus (TLCV)

Management

i) Use of certified seeds

ii) Growing of resistant varieties.

Minor diseases

Damping off *Pythium butleri*

Root rot *Acremonium* sp

Leaf spot *Alternaria alternata*

Leaf spot *Cercospora crotolariae*

Leaf spot *Helminthosporium crotolaria*

Leaf spot *Phoma glomerata*

Leaf spot *Phyllosticta crotolaria*

Leaf blight *Choenophora cucurbitarum*

Stem and root *Pellicularia rolfsii*

Charcoal *Macrophomina phaseolina*

Rust *Uromyces* sp

Viral diseases

Mosaic White clover mosaic virus

MESTA (*Hibiscus cannabinus*)

Fungal diseases

1) Anthracnose

Symptoms: The disease causes withering of plants.

Causal organism: *Colletotrichum hibisci*

Management

i) Spray Copper oxychloride 50%WP @3g/litre

2) Tip rot

Causal organism: *Phoma* sp

3) Root rot

Causal organism: *Rhizoctonia bataticola*

Minor diseases

Fungal diseases

Anthracnose *Colletotrichum hibisci*

Dry rot or Charcoal rot *Macrophomina phaseolina*

Collar rot and wilt *Phytophthora parasitica* var. *necotiane*

Root and stem disease *Pythium perniciosum*

Leaf and stem rot *Sclerotium rolfsii*

Grey mold *Botrytis* sp

Leaf spot *Cercospora abelmoschi*

Leaf spot *Cercospora hibiscina*

Leaf spot *Cercospora malayensis*

Leaf spot *Cercospora abelmoschi- cannabini*

Leaf spot *Diplodia hibiscina* var. *subdariffae*

Leaf spot *Phoma subdariffae*

Leaf spot *Phyllosticta hibiscina*

Leaf and flower spot *Drechslera tetramera*

Powdery mildew *Leveillula taurica*

Leaf and stem rot *Sclerotium rolfsii*

Target spot *Rhizoctonia solani*

Bacterial diseases

Bacterial wilt *Ralstonia solanacearum*

Viral diseases

Mosaic Mesta yellow vein mosaic virus (MYVMV)

6

Diseases of Sugar Crops

SUGARCANE (*Saccharum officinarum*)

Fungal diseases

1) Red rot

Symptoms: It is one of the most destructive disease of sugarcane. Dark reddish spot with brown margin appear along the mid rib. When the diseased canes are split open red blotches with transversely elongated white centres are seen. Under humid conditions numerous black dot like acervuli develop in the central discoloured portions. The diseased tissues emit a typical alcoholic odour. The juice of the affected stalks gives a sour alcoholic smell.

Causal organism: *Colletotrichum falcatum*

Management

i) Select healthy and disease free setts.

ii) Dip the setts on Carbendazim 50% WP or Thiophanate Methyl 70% WP @0.15% solution for 30 minutes before planting.

iii) Avoid rationing of infested crops.

iv) Growing resistant/tolerant varieties, i.e. Co6907, Co 7219, Co838, Co86249, Co87263.

v) Spray the crop with Carbendazim 50%WP (0.15%).

vi) Avoid water logging.

vii) Uproot the infested clump and destroy.

2) Smut

Symptoms: The inflorescence or the growing axis is replaced by a long whip like dusty black shoots. The smut powder on this whip is covered by a thin white silvery membrane made up of host epidermis. It ruptures soon exposing dense black powdery mass consisting of the smut spores. The spores can be blown about by even a gentle breeze. The affected canes are slender and thinner. In the infected canes if growing apex is damaged as a result of insect or mechanical injury before the whip is produced. Several eyes on the cane can sprout and invariably produce smut whips.

Causal organism: *Ustilago scitaminae*

Management

i) Smutted whips carefully collected in a bag and buried them in deep soil.

ii) Avoid ratooning of infested crops.

iii) Growing resistant /tolerant varieties, i.e. Co6907, Co87263

iv) Disinfect the canes with Carboxin (37. 5%)WP + Thiram (37.5%) WP @ 0.2%

3) Sugarcane wilt

Symptoms: The symptom appears late in the season as yellowing and withering of crown leaves followed by rapid drying of the cane.The internal tissue of the cane becomes brown or reddish brown in colour but there is no transverse white bands which is seen only in red rot affected canes If the infected canes are splitted they emit unpleasant odour.

Causal organism: *Cephalosporium sacchari*

Management

i) Select canes from completely healthy plots

ii) All the infested canes together with their leaves should be collected and burnt.

iii) No ratoon should be grown in the infested fields.

iv) Sett treatment with boron and zinc at the rate of 40ppm has been reported to reduce the wilt infestation.

4) Eye spot

Symptoms: The initial symptoms develop as elongated pale yellow spots on both sides of the young leaves. Later on the spots turn red with yellow margin. The leaf becomes brown.

Causal organism: *Drechslera sacchari*

Management

i) Growing of resistant variety

ii) Application of balance dose of nitrogen

5) Rust

Symptoms: Yellowish brown spots develop on both the sides of leaves and leaf sheaths. Numerous lesions coalesce cause dying of leaves.

Causal organism: *Puccinia melanocephala*

Management

i) Grow resistant varieties

ii) Spraying of Chlorothalonil 75%WP (0.2%) twice or thrice

iii) Removal of debris after harvesting of the crop.

6) Yellow spot

Symptoms: The irregular yellow green spots appear on leaves and leaf sheaths. These spots may increase in size and coalesce. In severe infection the entire field reveals a rusty yellow appearance. The leaves shrivel and die prematurely.

Causal organism: *Cercospora kopkeii*

Management

i) Spraying of Copper oxychloride 50% WP @3g/litre after appearance of the disease in the field

7) Root rot

Symptoms: The root system of affected plants are much reduced with few smaller and feeder roots. Plants may show wilting symptom. Tillering is delayed and reduced.

Causal organism*: Pythium graminicola*

Management

i) Avoid waterlogged condition

ii) Balance dose of nitrogen application

iii) Growing of resistant varieties like CO-290 and Kassoer

8) Pine apple disease

Symptoms: Cuttings may decay before germination or the shoots may show die back symptom. The odour of rotting resembles like pine apple odour.

Causal organism: *Ceratocystis paradoxa*

Management

i) Planting of healthy setts

ii) Treating of setts with Mancozeb 50% WP (0.3%)

iii) Light irrigation isto be given after planting till the complete of germination.

iv) Maintaining of good drainage in the soil.

9) Downy mildew

Symptoms: The typical symptom is the development of chlorotic stripes on the leaves and becomes more prominent as the leaves elongate. The stripes are light green and become dark red as the leaves become older in age. The stalks are light, brittle and worthless for milling.

Causal organism: *Sclerospora sacchari*

Management

i) Growing of resistant varieties

ii) Treating of setts with Mancozeb 50%W P @0.3% for 10 minutes

iii) Diseased stalks should be harvested early

iv) Destruction of any volunteer plants

v) Rogue out of diseased stools in the field

10) Brown spot

Symptoms: Numerous reddish- brown, oval to linear spots appear on both surfaces of older leaves. Later the spots become broad and oval with ashy or straw coloured centres. The severely infected leaves may die prematurely giving affected plants and the entire fields a blighted appearance.

Causal organism: *Cercospora longipes*

Management

i) Growing of resistant varieties

ii) Treating of setts with Mancozeb 50%WP (0.3%) for 10 minutes

11) Red Sheath

Symptoms: Both the sides of the sheath discoloured become reddish coloured. The stalk, rind and buds when covered the diseased sheath are also affected by this disease. The taste of the canes also affected.

Causal organism: *Sclerotium rolfsii*

Management

i) Growing of resistant varieties

ii) Treating of setts with Mancozeb 50% WP(0.3%) for 10 minutes before planting

12) Sooty Mould

Symptoms: A dark brown to black film or crust of fungal growth appear on leaves and leaf sheaths. They grow on the secretions of the insect

Causal organism: *Capnodium* sp

Management

i) Growing of resistant varieties

ii) Treating of setts with Mancozeb 50% WP (0.3%) for 10 minutes before planting

iii) Spraying of setts with 1% Starch solution.

13) Rind or Sour Rot disease

Symptoms: Irregular, dark, discoloured areas appear on stalk. The pith portion rots. Numerous black, coiled hair like structures appear on the rind. The internal tissues of the stalk become dark colour and have sour smell. The leaves become yellow, dry up and covered with black coiled structures.

Causal organism: *Phaeocytostroma sacchari*

Management

i) Harvesting of canes at proper time.

14) Slerophthora disease

Symptoms: The affected plants show severe stunting. The leaves of the affected plants become small and brittle. Chlorotic patches observed on leaves giving mosaic appearance

Causal organism: *Sclerophthora macrospora*

Management

i) Growing of resistant varieties

ii) Treating of setts with Mancozeb 50%WP (0.3%) for 10 minutes before planting

iii) Clean cultivation

iv) Removal of wild grasses from the original field and also adjacent fields.

Bacterial diseases

1) Red stripe and Top Rot

Symptoms

Red stripe: The disease consists of two phases leaf stripes and top rot. Long narrow dark red sripes develop on the leaves of young canes and ratoons. Early infection appears commonly on young and middle leaves as light green stripes often near the mid ribs that rapidly extend on both the directions and assume red colouration.

Top Rot: In this disease plant exhibit wilting and yellowing of older leaves. Bud infection can occur directly without showing leaf symptoms and the affected internodes exhibit sunken water-soaked areas that are brown to red in colour. The terminal bud die finally.

Causal organism*: Xanthomonas rubrilineans*

Management

i) Planting of healthy setts of resistant varieties.

ii) Treating of setts with Copper oxychloride @0.2% for 10 minutes.

iii) Do not select cane setts from the diseased field for planting.

2) Gummosis

Symptoms: The pale yellow to brown longitudinal stripes towards the tip of the leaves. The internodes are thin with numerous shoots on lower nodes resulting into bushy appearance. When the affected cane is the pith is shown to be water-soaked and yellow gummy ooze.

Causal organism: *Xanthomonas axonopodis* pv. *vasculorum*

Management

i) Planting of healthy setts of resistant varieties.

ii) Treating of setts with Copper oxychloride @0.2% for 10 minutes.

iii) Do not select cane setts from the diseased field for planting.

Phytoplasma diseases

1) Grassy shoot or Albino disease

Symptoms: The disease is characterized by the production of numerous thin tillers from the base of the affected stool which gives the appearance of grassy stool with pale yellow or chlorotic and small thin narrow leaves. Shoot development is poor with short internodes

Causal organism: Phytoplasma

Management

i) Use of setts from healthy plots

ii) Do not ratoon the diseased plants

iii) Hot air treatment of seed canes at 54°C for eight hours

Bacterial diseases

1) Ratoon stunting

Symptoms: The infected plants show stunted growth with short and thin cane. The characteristic symptoms can be seen when the cane is splited into two halves there is pink discolouration of growing point. In matured canes the vascular bundles in the nodes show yellow-orange, pink red or reddish brown. The ratoon crops are more severely infected.

Causal organism: *Leifsonia xyli* sub sp. *xyli*

Management

i) Use of setts from healthy plots

ii) Do not ratoon the diseased plants

iii) Hot air treatment of seed canes at 54°C for eight hours

Viral diseases

1) Mosaic

Symptoms: The symptoms are more pronounced on the newly emerged leaves. Pale yellowish patches appear in green tissues of leaves. These are irregular in shape and size.

Causal organism: Sugarcane mosaic virus (SCMV)

Management

i) Planting of healthy setts of resistant varieties. Neela Madaham, Neela Chakra, Charchika and Raghunath.

ii) Do not select cane setts from the diseased field for planting.

iii) Rouging out of infected plants

Root parasites

1) Bunga

Symptoms: It is parasite on sugarcane roots. It is of 3cm height with rudimentary scaly leaves and purplish black flowers.

Causal organism: *Aeginetia indica*

Management

i) Removal of stools along with the parasite

ii) Ploughing of infested field.

2) Striga

Symptoms: It is parasite on sugarcane roots. It is of 30 to 45cm tall height with narrow green leaves and white flowers.

Causal organism: *Striga densiflora*

Management

i) Removal of stools along with the parasite

ii) Ploughing of infested field.

iii) Growing of resistant varieties

ii) Treating of setts with 0.3% Mancozeb for 10 minutes before planting

Non-parasitic diseases

1) Yellowing

Symptoms: Leaves turn yellow from tip downwards. Later on the whole plant dry up.

Causal factor: Physiological deficiency

Management

i) Application of balance dose of fertilizer

Minor diseases

Fungal diseases

Alternaria blight: *Alternaria alternata*

Banded Sclerotial disease: *Corticium Sasakii*

Cephalosporium wilt: *Cephalosporium sacchari*

Leaf spot: *Phoma sorghina*

Leaf spot: *Phyllosticta sacchari*

Ergot : *Claviceps purpurea*

Fusarial sett and stalk rot: *Giberella fujikuroi*

Leaf spot: *Curvularia lunata*

Lea spot: *Phoma sorghina*

Necrotic leaf spot: *Curvularia tuberculata*

Seedling blight: *Drechslera halodes*

Bacterial diseases

Leaf scald: *Xanthomonas albilineans*

Viral diseases

Mosaic: i) Chlorotic streak virus, ii) Striate mosaic virus

7

Diseases of Medicinal Crops

MINT (*Mentha* sp)

Fungal diseases

1. Stolon Rot

Symptoms: Yellowing of leaves is observed at early stage of infection. In later stage wilting of plants are noticed followed by drying of the entire plants.

Causal organism: *Macrophomina phaseoli*

Management

i) Stolons should be collected from healthy fields.

ii) Use of healthy and disease free stolons for planting.

ii) Treatment of stolons with Metalaxyl 8% + Mancozeb 64% WP (0.3%) before planting.

2. Rust

Symptoms. In the spring or summer season the leaves and upper portion of stems are affected and become twisted. Yellow/ dusty orange pustules appear on the leaves and stems. In later part the pustules become black. Drying of leaves followed by defoliation occur.

Causal organism: *Puccinia menthae*

Survival and spread: From infected plant the spores are spread to healthy plants by wind in the month of June and July when the pustules mature and break open.

Favourable conditions

i) Dry weather with high relative humidity

ii) Pathogens perpetuate on wild mint plants and left over stubbles in the field

iii) The pathogen is wind borne

iv) It spreads through diseased propagating material.

Management

i) Field sanitation

ii) Crop rotation with non-host crops

iii) Use of disease free planting materials

iv) Planting of stolons at 5cm depth.

v) Spraying of crop with Copper oxychloride 50% WP (0.3%) /Mancozeb 75% WP (0.3%) / Hexaconazole (0.1%) two to three times at 7 days interval after appearance of the disease in the field.

3. Alternaria Leaf Blight

Symptoms. Small brown spots appear on the upper surface of the leaves and gradually increase in size and these are surrounded by yellow halo. In advanced stage the whole leaf is covered with spots and show drying followed by defoliation.

Causal organism: *Alternaria* sp

Management

i) Clean cultivation and removal of crop debris from the field

ii) Planting of disease free stolons collected from healthy fields.

iii) Spraying of crop with Mancozeb 75% WP (0.3%) two to three times at 7 days interval after appearance of the disease in the field.

4. Verticillium Wilt

Symptoms. The top leaves becomes yellowing show drooping and the growth of the plants are checked. Premature defoliation and death of the plant may occur.

Causal organism: *Verticillium dahliae*

Survival and spread: The pathogen is soil borne. Infected planting materials help in spread of the disease.

Management

i) Field sanitation

ii) Crop rotation for three years with non-host crops.

iii) Deep summer ploughing. Removal and destruction of diseased plant debris

iv) Timely planting

v) Treatment of planting materials with *Trichoderma* sp @4g/kg.

5. Black Stem Rot

Symptoms. Dark brown to black cankers appear on the stems. In severe infection these cankers may girdle the stem and cause the death of the above plant parts. The pathogen become active in cool and wet weather

Causal organism: *Phoma strasseri*

Survival and spread: The pathogen survive in soil and also infected plant debris by producing 1to 3 mm diameter sized brown to black sclerotia.

Favourable conditions: The disease become more severe in moderately wet soil in 15 to 25 °C

Management

i) Field sanitation

ii) Crop rotation for three years

iii) Deep summer ploughing. Removal and destruction of diseased plant debris

iv) Timely planting

v) Treatment of planting materials with *Trichoderma* sp @4g/kg before planting.

vi) Spray the foliar fungicides like Difenconazole 25EC@ 0.1% or Validamycin 3% L (0.2%) two to three times at 7-10 days interval after occurrence of the disease in the field.

6. Septoria leaf spot

Symptoms. Dark brown to black colour up to 3mm size angular spots appear on leaves.

Causal organism: *Septoria lycopersici*

Survival and spread

The fungus survives in plant debris and on weeds.

Favourable conditions

Hot weather and intermittent rains favour disease development.

Management

i) Summer ploughing
ii) Clean cultivation
iii) Use of certified planting materials.
iv) Growing of resistant /tolerant varieties for different categories of mint crops developed from Central Institute of Medicinal and Aromatic Plants, Lucknow, Uttar Pradesh)
    a) Japanese mint (*Mentha arvensis*): CIMAP-MAS-1, CIMAP-Hybrid-77, Shivalik, EC-41911 Himalaya
    b) Spear mint (*M. spicata*): CIMAP-MSS-1,CIMAP-MSS-5 and CIMAP-MSS-98, Punjab, Arka, Neera
    c) Beramot mint (*M. citrata*): Kiran
    d) Pepper mint (*M. piperita*): Kukrail, Pranjal, Tushar

7. Powdery mildew

Symptoms: Powdery patches appear on the leaves and other green parts. Later on these patches become dull coloured are covered with black dot like cleistothecia. The leaves become dry and fall down.

Causal organism: *Erysiphe cichoracearum*

Survival and spread

Primary infection is through the resting spores and secondary spread occurs through infected planting materials and wind.

Management

i) Field sanitation

ii) Crop rotation

iii) Spraying of crop with Sulphur 80% WP @ 4gms/litre two to three times at 15 days interval

iv) Application of judicious dose of Nitrogen and Phosphorus fertilizers.

v) Growing of resistant /tolerant varieties: There are four types of mint crops, i.e. (a) Japanese mint (b) Spear mint (c) Beramot mint (d) Pepper mint

a) Japanese mint: CIMAP-MAS-1, CIMAP-Hybrid-77, Shivalik, EC-41911

b) Spear mint: CIMAP-MSS-1,CIMAP-MSS-5 and CIMAP-MSS-98, Punjab, Arka, Neera

c) Beramot mint: Kiran

d) Pepper mint: Kukrail, Pranjal,Tushar

8. Anthracnose (leopard spot).

Symptoms. Circular, oval sunken spots appear on leaves. These spots show reddish brown margin with light grey centre. These spots enlarge and coalesce with each other causing blightening of leaves. In severe infection defoliation may occur. The lesions may appear on the stem causing splitting in later stage.

Causal organism: *Sphaceloma menthae*

Survival and spread

The pathogens survive in infected plant parts in spring, and also throughout summer.

Favourable condition

Leaf wetness for 3 to12 hours.

Mangement

i) Field sanitation

ii) Deep summer ploughing. Removal and destruction of diseased plant debris

iii) Timely planting and fertilizer application WP@0.15%

iv) Spraying of crop with Mancozeb 75% WP @3 gm/litre two to three times at 7 days interval.

Fungal diseases

Minor diseases

Leaf spot: *Alternaria alternata*

ASWAGANDHA (*Withania somnifera*)

1) Damping off

Symptoms: Water-soaked necrotic lesions are observed on base of the stem which ultimately cause death of the plant.

Causal organism: *Rhizoctonia solani*

Management

i) Field sanitation

ii) Crop rotation with non-host crops for three years

iii) Deep summer ploughing. Soil solarisation. Removal and destruction of diseased plant debris

iv) Seed treatment with both Carbendazim 12% + Mancozeb 63%WP @ (0.2%) *T. viride* (10g/kg) and as application of neem cake@500gm/sq. mtr at the time land preparation.

2) Root rot

Symptoms: It is an important soil borne disease causing significant yield losses. The roots of infected plant become brown colour and pulpy. Cracks develop on old lesions and cortical tissues become discoloured and decayed. The infected plants show white cottony growth near ground level. Plants show severe wilting resulting death at seedling stage cause 30 to 50% mortality. If the disease occurs at later stage plants show withering and drooping.

Causal organism: *Fusarium solani*

Management

i) Field sanitation

i) Deep summer ploughing. Soil solarisation. Removal and destruction of diseased plant debris.

iii) Application of judicious dose of Nitrogen and Phosphorus fertilizers.

iv) Seed treatment with both Carbendazim 12% +Mancozeb 63% WP @ (0.2%) and *T. viride* (10g/kg) and application of neem cake @500gm/sq. mtr before planting should be practised for management of the disease effectively

v) Soil application of Neem cake increases the percentage of germination and mortality of seedlings reduced. Seed treatment with *T. viride* reduces the disease incidence.

3) Leaf spot/Leaf blight

Symptoms: The small light brown to brown spots appear on both the surface of leaves. Gradually these spots increas in size. Several spots coalesce becoming irregular, dark brown with concentric rings and cause blightening of leaves. In severe infection 80-90% leaves are damaged. Such leaves often fall of prematurely.

Causal organism: *Alternaria alternata*

Management

i) Field sanitation

ii) Seed treatment.

iii) Deep summer ploughing. Removal and destruction of diseased plant debris

iv) Spraying of crop with Mancozeb 75% WP (0.3 %) two to three times at 7 days interval after appearance of the disease in the field.

Minor diseases

Fungal diseases

Anthracnose: *Colletotrichum capsici Colletotrichum gloeospoiriodes*

Curvularia leaf spot: *Curvularia lunata*

Leaf spot: *Alternaria alternata, Cercospora serpentinae, Mycospharella rauwolfiae, Pellicularia filamentosa*

Powdery mildew: *Leveillula taurica*

Root rot: *Macrophomina phaseoli*

Target spot: *Corynespora cassicola*

ALOE VERA *(Aloe barbadensis)*

Fungal diseases

1) Rust

Symptoms: Pale yellow spots appear on the upper leaves which increase in size and become brown or black circular spots. The black spots are created by the oxidation of the phenols present in plant sap and are permanent once they appear. The infected leaves drop prematurely.

Causal organism: *Phakopsora pachyrhizi*

Host plants: Beans, Allium, corn, sorghum, rice, banana, coffee, and a wide range of agricultural crops

Affected plant stages: Vegetative stage

Affected plant parts: Stems, leaves, petioles, and pods

Favourable condition

i) Frequent rains, drizzle or dew with cool temperature and high humidity.

ii) When poor quality sseds are used in poor field conditions the disease become more severe.

Management

i) Field sanitation

ii) Seed treatment

iii) Deep summer ploughing

iv) Removal and destruction of diseased plant debris

v) Spraying of crop with Mancozeb 75% WP (0.3%) or Propiconazole 25% RC (0.1%) two to three times at 7 days interval after the appearance of the disease in the field.

2) Anthracnose

Symptoms. Small round to oval dark green water soaked spots about 1-2 mm in size appear on leaves which later becomes circular. The centre become tan to light brown. In severe infection both the surface of leaves infected ultimately lead to the death of infected leaves. The affected area lost the mucilaginous gel.

Causal organism *Colletotrichum gloeosporioides*

Favorable Condition

The disease is favoured by hot and humid condition. The pathogen spreads easily in rainy weather by water splash.

Management

i) Field sanitation
ii) Crop rotation
iii) Deep summer ploughing. Removal and destruction of diseased plant debris
iv) Use healthy disease free cuttings for planting

3) Basal stem rot

Symptoms: In cold and damp condition the base of the plant near soil level rot and become dark red or black. It is due to the exposure of the stems near the soil to excessively wet soil. As the tissues rot, it changes color from reddish brown to black. The color of the rot depends upon the type of the plant and the nature of the fungus attacking it. The plant leoks to be dying from the base upward.

Causal organism: *Fusarium* sp

Favourable condition

Disease is favoured by warm and wet weather and it spreads easily during wet weather by water splash

Management

i) Field sanitation
ii) Crop rotation
iii) Deep summer ploughing. Removal and destruction of diseased plant debris.
iv) Basal drenching of the crop with Carbendazim 50% WP (0.15%) two to three times at 7 days interval after appearance of the disease in the field.

4) Leaf spot

Symptoms: Circular to oval dark brown small spots appear on leaves mostly towards leaf tip. Spots enlarge became brown to black with age and were surrounded by a yellow halo. Later irregular to elongated, sunken lesion appear on both leaf surfaces causing blightening followed by premature drop off.

Causal organism: *Alternaria alternata*

Favourable condition

Disease is favoured by warm and wet wet weather and spread easily during wet weather by water splash.

Management

i) Field sanitation

ii) Crop rotation

iii) Deep summer ploughing. Removal and destruction of diseased plant debris

iv) Spraying of crop with Mancozeb 75% WP (0.3%) two to three times at 7 days interval after appearance of the disease in the field.

5) Sooty mold

Symptoms: Sooty mold is a fungal infection that is secondary after infestation of either aphids or mealy bugs. The aphids and mealy bugs secret honeydew like sticky substance on the leaves on which some fungus grow rapidly and produce masses of black spores which stick to the leaf surface. The foliage appears black and ugly. The black /dark coating block the light inhibit the effective area of photosynthesis and growth of the plant reduced.

Causal organism

i) *Cladosporium* sp

ii) *Aureobasidium* sp

ii) *Antennariella* sp

iii) *Limacinula* sp

iv) *Scories* sp

v) *Capnodium* sp

Management

i) Spray the crop with Diazinon (0.03%) followed by spraying with a starch solution. The starch coating dries up and come off in form cf flakes.

6) Root rot

Symptoms

The plants show stunted growth and yellowed leaves and the roots become rotted. The lower leaves become mushy. The roots turn dark brown and tips become mushy. In Pythium infestation when a plant is pulled the roots may remain in the soil.

Causal organism

i) *Pythium* sp

ii) *Phytophthora* sp.

Favourable Condition

Low temperature and excessive watering.

Management

i) Field sanitation

ii) Crop rotation

iii) Deep summer ploughing.

iv) Removal and destruction of diseased plant debris.

Bacterial diseases

1) Bacterial soft rot

Symptoms. Initially the outer petiole of the leaf which is in contact with the soil become watery and then gradually the symptom progresses. Then the whole leaf become watery and often has a complete leaf rot. The affected area becomes soft and mushy and generally turn dark in colour. An infected leave has a soft and watery decay. Infected leaves rots internally resulting in the wilting of young leaves which is then followed by collapse of tip and die back. The diseased plant often has a foul odour. The disease becomes serious when abundant moisture is available through irrigation or rain.

Causal organism: *Pectobacterium chrysanthemi*

Host plant: Potato, sweet potato, cassava, onion, cabbage, carrot, tomato, beans, cotton, coffee, banana.

Mode of spread: Primary and secondary spread are through water borne bacteria.

Favourable condition: The pathogen is soil borne and survive in plant debris in the field and the disease is favoured by hot and wet weather.

Management

i) Field sanitation. Removal and destruction of diseased plant debris

ii) Crop rotation with non-host crops for three years

iii) Deep summer ploughing.

iv) Judicious application of water to the plants.

v) Plants should be grown under adequate sunlight condition and keep water off the leaves.

vi) Proper spacing should maintained among the plants.

vii) Basal drenching of the crop with Copper oxychloride 50%WP (0.25%) two to three times at 7 days interval after appearance of the disease in the field.

LIQUORICE (*Glycyrrhiza glabra)*

1) Root rot and Collar rot

Symptoms: Yellowing and defoliation of leaves followed by drying of infected plants before maturity. Reddish brown necrotic lesions appeared on the roots more often advanced to the collar region of plants.

Causal organism: *Pythium* sp

Management

i) Field sanitation

ii) Crop rotation for three years

iii) Deep summer ploughing. Removal and destruction of diseased plant debris

iv) Timely planting

v) Treatment of planting materials with *Trichoderma* sp @4g/kg.

2) Stem rot

Symptoms: Web like growth develop at the base of the plant near soil level. Stem portion show rotting followed by death of the plant.

Causal organism: *Rhizoctonia* sp

Management

i) Field sanitation

ii) Crop rotation for three years

iii) Deep summer ploughing. Removal and destruction of diseased plant debris

iv) Timely planting

v) Treatment of planting materials with *Trichoderma* sp @4g/kg.

3) Damping off

There are two types of damping-off (i) Pre-emergence damping-off (ii) Post-emergence damping-off.

In pre-emergence damping-off the seeds may be rotted.

Symptoms: In post-emergence damping-off the plants are died after germination. There may be stunting of plants and rotting of roots.

Causal organism: *Pythium* sp

Management

i) Field sanitation

ii) Crop rotation for three years

iii) Deep summer ploughing. Removal and destruction of diseased plant debris

iv) Timely planting

v) Avoid excessive irrigations.

vi) Treatment of planting materials with *Trichoderma* sp @4g/kg.

4) Botrytis gray mold

Symptoms: Water-soaked spots appear on the leaves and stems which expand rapidly into large yellowish-green or irregular greyish brown lesions. Ultimately the plants collapse.

This disease occur on wet foliage during overcast conditions and in high humidity environments.

Causal organism: *Botrytis* sp

Management

i) Avoid excessive irrigations.

ii) Use of certified seeds

iii) Seed treatment with Carbendazim 50%WP (0.15%)

5) Alternaria leaf spot

Symptoms: Brown to dark brown spots appear on the leaves. These spots coalesce cause blightening of leaves. The disease is mostly confined to older leaves in adult plants but it may infect new leaves of young plants particularly in the rainy season.

Causal organism: *Alternaria* sp.

Management

i) Seed treatment with Carbendazim 50% WP (0.15%)

ii) Spraying of crop with Mancozeb 75% WP (0.3%) twice/thrice at 7 days interval after appearance of the disease in field.

PALMA ROSA (*Cymbopogon martinii*)

1) Red leaf spot

Symptoms. The disease is much prevalent in rainy season in all palmarosa growing regions. The disease appears as small brownish spots on leaf surface, spreads rapidly under warm and humid conditions and may cause 35% reduction in oil production. The disease also infects leaf sheaths and mid rib.

Causal organism: *Colletotrichum caudatum*

Favourable condition

High relative humidity (90-97%) and moderate temperature (24-28°C) favour the disease development

Management

i) Field sanitation.

ii) Deep summer ploughing. Removal and destruction of diseased leaves which fall on the ground.

ii) Timely planting

iv) Spraying of crop with Mancozeb 75%WP (0.3%) twice or thrice at 7 days interval after appearance of the disease in field.

2) Ellisiella blight

Symptoms: Small greyish spots appear on the surface of infected leaves during rainy season. These spots increase in size, coalesce and cause blightening of leaves. Ultimately the infected leaves dry up.

Causal organism: *Ellisiella caudate*

Management

i) Field sanitation

ii) Crop rotation for three years

iii) Deep summer ploughing. Removal and destruction of diseased plant debris

iv) Timely planting

v) Spraying of crop with Mancozeb 75% WP (0.3%) twice or thrice at 7 days interval after appearance of the disease in field.

3) Little leaf or Grassy shoot

Symptoms. In this disease both seed production and oil yield is reduced. The inflorescence is converted into leafy structure.

Causal organism: *Balansia* sp

Management

i) Field sanitation

ii) Crop rotation for three years

iii) Deep summer ploughing. Removal and destruction of diseased plant debris

iv) Seed treatment

v) Spraying of crop with Dithane Z-78(0.3%) or Mancozeb 75% WP (0.3%) just at the time of flowering.

4) Smut

Symptoms: The seed is converted into smut sori which are slightly covered by dull yellow membrane. Due to this disease the oil and seed yield reduce.

Causal organism: *Tolyposporium christensenii*

Management

i) Field sanitation

ii) Crop rotation for three years

iii) Deep summer ploughing. Removal and destruction of diseased plant debris

iv) Timely planting

v) Spraying of crop with Dithane Z-78 (0.3%) or Mancozeb 75% WP (0.3%) just at the time of flowering.

6) SARPGANDHA (*Rauwolfia serpentina)*

1) Cercospora leaf spot

Symptoms. Dark brown coloured spots appear on the upper surface of the leaves. The corresponding lower surface shows yellowish brown. Infected leaves turn yellow become dry and fall off.

Causal organism: *Cercospora rauwolfia*

Epidemiology: Primary source of inoculum is the infected host and secondary spread occurs by air. Temperature (21-23 °C) and RH (80%) favours disease spread.

Management

i) Field sanitation

ii) Crop rotation with non –host crops

iii) Removal and destruction of diseased plant debris

iv) Timely planting

v) Treatment of seed with Carbendazim 50% WP (0.15%).

vi) Spraying of crop with Mancozeb 75% WP (0.3%) twice or thrice at 7 days interval after appearance of the disease in field. Preventing spraying can be given before monsoon.

2) Cercospora Leaf Blotch

Symptoms. Purple mud coloured small dots appear on the ventral surface of leaf which gradually increase in size and form circular patches. Later these patches coalesce to an irregular form. Within 7 days the affected portion turns maroon and the whole leaf dies.

Causal organism: *Cercospora serpentinae*

Epidemiology: Primary source of inoculum is infected host and secondary spread occurs by air. Temperature (21-23 °C) and RH (80%) favours disease spread.

Management

i) Field sanitation

ii) Crop rotation with non –host crops

iii) Removal and destruction of diseased plant debris

iv) Timely planting

v) Treatment of planting materials with Carbendazim 50% WP (0.15%).

vi) Excess irrigation is avoided

vii) Close planting is avoided. Adequate spacing is maintained.

viii) Spraying of crop with Mancozeb 75%WP (0.3%) twice or thrice at 7 days interval after appearance of the disease in field. Preventing spraying can be given before monsoon.

3) Alternaria leaf spot

Symptoms: Small brown coloured spots appear on the leaves. These spots increase in size and coalesce with each other and cover the entire leaf surface. The infected leaves become blighted. In severe infection flowers and fruits are affected. In stem violet black lesions appear.

Caused Organism: *Alternaria tenuis*

Management

i) Field sanitation

ii) Crop rotation

iii) Removal and destruction of diseased plant debris

iv) Timely planting

v) Seed treatment with Carbendazim 50% WP (0.15%)

vii) Spraying of crop with Mancozeb 75% WP (0.3%) twice or thrice at 7 days interval after appearance of the disease in field.

4) Die back

Symptoms. The affected plants strat drying from tip downwards.

Causal organism: *Colletotrichum dematium*

Favourable condition

High relative humidity (90-97%) and moderate temperature (24-28°C) favours the disease development

Management

i) Field sanitation

ii) Affected branches may be pruned

iii) Removal and destruction of diseased plant debris

v) Spraying of crop with Mancozeb 75% WP (0.3%) twice at 7 days interval after pruning.

5) Top Dying

Symptoms: The disease usually appears in the field after rains during the September. Initially brown spots appear on the leaf lamina which later coalesce to form larger spots up to 5 to 11 mm in size. Small pin head like dark blackish brown pycnidia develop on the leaf lamina. The tender shoots first got infected by the pathogen later the infection moved downwards causing death of the plant. The severe infection on twigs and leaves also lead to premature drying and death of the entire plant.

Causal organism: *Lasiodiplodia theobromae*

Management

i) Field sanitation

ii) Affected branches may be pruned

iii) Removal and destruction of diseased plant debris

v) Spraying of crop with Mancozeb 75% WP (0.3%) twice at 7 days interval after pruning.

6) Root Rot

Symptoms: The leaves become yellow and show drooping followed by blackening of stem, rotting of the root and base portion of the stem as well as peeling of the stem, bark and root epidermis. The rotted portion shows the presence of black sclerotia.

Causal organism: *Macrophomina phaseolina*

Epidemiology: High humidity and poor air circulation promote disease development.

Management

i) Field sanitation

ii) Crop rotation

iii) Removal and destruction of diseased plant debris

iv) Timely planting

v) Spraying of % Mancozeb75WP @ 0.3%

vi) Soil should be well drained to prevent disease incidence.

7) Target Leaf Spot

Symptoms: The pathogen causes dark brown spots on upper and yellowish brown spots on the lower surface of the leaf. These spots gradually enlarge into circular spots 2-20 mm (dia) with concentric zones so called as target spot. Heavily infested leaves turn yellow and shed prematurely.

Causal organism: *Corynespora cassicola*

Epidemiology: Conidia disseminated by wind and rain where as chlamydospores serve as important survival structure. Temp (20-25°C) and RH (85%) favours the disease development. Continuous leaf wetness is the most important factor for spores to germinate.

Management

i) Field sanitation
ii) Crop rotation
iii) Removal and destruction of diseased plant debris
iv) Timely planting
v) Spray with Mancozeb before monsoon at monthly intervals.
vi) Removal of infected crop debris and weeds.

8) Fusarial wilt

Symptom: Infected plants remain stunted. The lower leaves turn pale green to golden yellow. The plants show droop off progressively upward from the base of the stem. Dark streaks may occur in vascular tissues of the roots and lower stem. The roots may decay. Infected plants wilt and die.

Causal organism: *Fusarium* sp.

Epidemiology: Primary source of inoculum is soil borne further spread occurs by contaminated implements. Temp (24°C) and RH (80%) favours disease spread.

Management

i) Field sanitation
ii) Crop rotation with non-host crops
iii) Seed treatment with Carbendazim 50% WP (0.15%)
iv) Removal of infected plant and destruction of diseased plant debris
v) Basal drenching with Carbendazim 50% WP (0.15%) twice or thrice at 10 days interval if necessary.
vi) Avoid over watering to reduce moisture.

9) Anthracnose

Symptoms. Numerous tiny spots scattered all over mainly on the upper surface of the leaf. The spot enlarges into circular patch and invades surrounding tissues. The spots coalesce causes drying of lamina resulting in defoliation.

Causal organism: *Colletotrichum gloeosporioides*

Epidemiology*:* Humid weather favours disease development. Leaf wetness directly related to disease severity. Temp of 27°C and RH of 80% helps its spread.

Management

i) Cut and burn the infected shoots

ii) Avoid overhead watering.

iii) Use Mancozeb 75% WP (0.3%) to control the disease.

10) Inflorescence & fruit rot

Symptoms: Initially water soaked lesions appear on the flowers and flower buds appear which later show slimy wet rot symptoms on the inflorescence and peduncles. The slimy wet rot moves both upward and downward and eventually cause premature death and drying of infected leaves. Fluffy cotton mycelial growth visible on infected inflorescences and fruits under high humidity conditions.

Causal organism: *Rhizopus stolonifer*

Epidemiology: High humidity favours disease spread.

Management

i) Field sanitation

ii) Crop rotation

iii) Removal and destruction of diseased plant debris

iv) Timely planting

v) Spraying of crop with Mancozeb75WP @ 0.3% after appearance of the disease twice or thrice.

Viral diseases

1) Mosaic

Symptoms: The primary symptom comprises vein banding and gradual yellowing of leaves. Later the leaf curls from the margin to the midrib on the abaxial surface and ultimately the leaf drops off.

Causal organism: Cucumber Mosaic Virus

Epidemiology: Hot weather and high humidity favours disease development.

Management

i) Field sanitation

ii) Crop rotation

iii) Removal and destruction of diseased plant debris

iv) Timely planting

v) Spraying of insecticides i.e. Imidacloprid 17.8% SL @ 1ml/3 litre or Thiamethoxam 25%WG 1.0g/5 litre,

vi) Select healthy planting material

vii) Remove weeds and infected crop residues to control the virus.

Minor diseases

Fungal diseases

Leaf spot: *Alternaria alternata*

Leaf spot: *Curvularia lunata*

Target spot: *Corynespora cassiicola*

Leaf spot: *Mycosphaerella rauwolfiae*

Leaf spot: *Pellicularia filamentosa*

SACRED BASIL (*Ocimum basilicum*)

Fungal diseases

1) Damping-off

Symptoms: Damping-off. There are two types of damping-off.

i) Pre-emergence damping-off

ii) Post-emergence damping-off

Pre-emergence damping-off: Death occurs before germination. Seeds fail to germinate.

Post-emergence damping-off: Death occurs after germination. Germinated seedlings collapse. Brown, shrivelled lesions appear at base of stem cause the toppling of the seedlings. Roots become brown and water-soaked cause death of the seedlings.

Causal organism

i) *Pythium ultimum*

Favourable condition

i) Promoted by high temperature and high humidity

ii) Poor air circulation.

Management

i) Field sanitation

ii) Plant seeds in sterile soil

iii) Plant basil seedlings in well-drained soil

iv) Seed treatment with Carbendazim 50%WP @ 0.3%

v) Removal of infected plant and destruction of diseased plant debris

2) Root rot

Symptoms: Roots become brown and water-soaked. Rotting of roots cause death of the seedlings.

Causal organism: i) *Rhizoctonia solani*

Favourable condition

i) Promoted by high humidity

ii) Poor air circulation.

Management

i) Field sanitation

ii) Plant seeds in sterile soil;

iii) Plant basil sedlings in well-drained soils.

iv) Seed treatment with Carbendazim 50%WP (0.15%)

v) Removal of infected plant and destruction of diseased plant debris

vi) Basal drenching with Carbendazim 50%WP (0.15%) twice or thrice at 10 days interval if necessary.

vii) Avoid over watering to reduce moisture.

3) Downy mildew

Symptoms: As the disease progresses, the most characteristic symptom of the disease i.e., fuzzy grayish- purple sporangia (the reproductive structures) develop on the lower surface of the leaves. Brown to black angular necrotic patches appear on the upper surface of the leaves.

Causal organism: *Peronospora belbahrii*

Management

i) Grow tolerant varieties

ii) Apply protective fungicide

iii) Ensure good air circulation around greenhouse grown plants

iv) Use drip irrigation to avoid wetting foliage

v) In the field or garden, good sanitation practices should be used to minimize inoculum levels.

4) Fusarial wilt

Symptoms: The plants become stunted. Leaves turn yellow. Brown spots or streaks appear on the stem. In severe infection stems become twisted. Suddenly leaves drop.

Causal organism: *Fusarium oxysporum* f. sp. *basilicum*

Disease cycle: The pathogen is soil borne. The disease may introduced into the field through by the use of contaminated seeds and infected plants as planting materials. Once introduced it can over winter through infected debris and cause infections to newly grown plants

Favourable condition

i) Promoted by high humidity

ii) Poor air circulation.

Management

i) Field sanitation

ii) Crop rotation with non-host crops

iii) Use of healthy and disease free seed.

iv) Seed treatment with Carbendazim 50% WP

v) Removal of infected plant and destruction of diseased plant debris

vi) Basal drenching with Carbendazim 50% WP (0.15%) twice or thrice at 10 days interval if necessary.

5) Cercospora leaf spot

Symptoms: Circular to irregular dark spots appear on the leaves with light centre. Several spots coalesce cause blightening of the leaves.

Causal organism: *Cercospora ocimicola*

Favourable condition

i) High humidity and poor air circulation favour the disease development

Management

i) Avoid overhead irrigation and splashing plants with water.

ii) Apply a layer of mulch around the plants to reduce water splash.

iii) Remove and destroy any symptomatic leaves.

iv) Minor infections can be controlled by spraying weekly with a fungicide containing potassium bicarbonate.

6) Grey mold

Symptoms. Brown to grey dense fungal growth appear on leaves and stems. Diseased leaves die and eventually fall from the plant. If severe lesions develop on main stem, the infected plant may die.

Causal organism: *Botrytis cinerea*

Favourable condition

i) High humidity and poor air circulation favour the disease development

ii) Moderately cool temperatures

Management

i) Avoid working in field in rainy conditions.

ii) Remove infested leaves and or plants.

iii) Avoid overhead irrigation.

Bacterial diseases

1) Bacterial Leaf spot

Symptoms: Angular or irregular brown or black water-soaked spots appear on leaves. In severe cases streaks occur on stems. The disease might not be severe in field conditions but it can cause damage to the seedlings.

Causal organism: *Pseudomonas cichorii*

Favourable condition

High humidity and overhead watering promotes spread of disease.

Management

i) Use disease free seed

ii) Use wide field spacing to promote air circulation around plants

iii) Removal of infected plants/ diseased leaves from plant and soil surface immediately.

PLANTAGO (*Plantago ovata*)

Fungal diseases

1) Damping off

i) Pre-emergence damping-off

ii) Post-emergence damping-off

Pre-emergence damping-off: Seeds could not be able to germinate. Death occurs before germination

Post-emergence damping-off: Death of the seedlings occurs after germination.

Initially water-soaked spots appear in collar region of the seedlings and the infection extends rapidly leading to death of the seedlings. The disease naturally occurs at higher temperature and high humidity at the time of sowing.

Causal organism

i) *Pythium ultimum*

ii) *Rhizoctonia solani*

iii) *Fusarium oxysporum*

Favourable condition

i) Promoted by high humidity ii) Poor air circulation.

Management

i) Use healthy plants from disease free fields.

ii) Pre-sowing seed treatment with Thiophanate methyl @ 2–3 gram/kg

iii) Avoid over irrigation.

iv) Spraying of Carbendazim 50%WP (0.3%) at 15 days after sowing.

2) Downy mildew

Symptoms: This disease usually appears at the time of spike initiation. Ashy white mycelial growth develop on the lower surface of the leaves while the corresponding upper surface become chlorotic and brownish in colour followed by drying of plants. In some cases chlorotic streaks extend along the mid rib of leaves in which profuse downy growth is observed. Finaly the whole leaf turns necrotic and the entire plants become blighted.

Causal organism

i) *Peronospora alta*

ii) *P. plantaginis*

Favourable condition

i) Promoted by high humidity

ii) Poor air circulation.

Management

i) Use of healthy disease free seed

ii) Seed treatment with Metalaxyl 8% WP + Mancozeb 64%WP at the rate of 2–3 g/kg

iii) Maintain proper spacing

iv) Avoid over irrigation.

v) Spray Metalaxyl 8% + Mancozeb 64% WP (0.3%)after appearance of the disease. If necessary spraying is repeated for twice/thrice at 7 days interval.

3) Leaf blight

Symptoms: The older and matured leaves are affected first. Small brown spots appear on the leaves which rapidly enlarges cause blightening of leaves. The leaf tips become dull green yellowish colour.The shrevelling and blackening of seeds may occur in severe infection.

Causal organism: *Alternaria alternata*

Favourable condition

i) Promoted by high humidity

ii) Poor air circulation.

Management

i) Field sanitation

ii) Crop rotation

iii) Removal and destruction of diseased plant debris

iv) Timely planting

v) Spraying of crop with Mancozeb 50% WS@ (0.2%) twice or thrice at 7 days interval after appearance of the disease in field.

4) Powdery mildew

Symptoms: The whitish or greyish small spots appear on the leaves which gradually enlarge and cover the entire leaf surface. In severe infection young twigs and flowers are affected.

Causal organism: *Erysiphe cichoracearum*

Favourable condition

i) Promoted by high humidity

ii) Poor air circulation.

Management

i) Sanitation

ii) Crop rotation

iii) Removal and destruction of diseased plant debris

iv) Timely planting

v) Spraying of crop with Karathane (0.15%) or Wettable sulphur compounds (0.4%) twice or thrice at 7 days interval after appearance of the disease in field.

5) Wilt

Symptoms: Lower leaves show change in normal colour. Fungal growth block the xylem vessels. Vascular discolouration is a common symptom. Discolouration occurs on the tap root. Gradually the plants show wilting.

Causal organism

i) *Fusarium oxysporum*

ii) *F. solani*

Favourable condition

i) Promoted by high humidity

ii) Poor air circulation.

Management

i) Field sanitation

ii) Crop rotation

iii) Deep summer ploughing. Removal and destruction of diseased plant debris.

iv) Use of healthy certified seed.

iv) Seed treatment with Carbendazim 50% WP (0.25%)

iv) Basal drenching of the crop with Carbendazim 50 % WS (0.25%) two to three times at 7 days interval after appearance of the disease in the field.

v) Avoid over-irrigation.

Minor diseases

Fungal diseases

Leaf spot: *Septoriae plantaginae*

PYRETHRUM (*Chrysanthemum cineriifolium*)

Fungal diseases

1) Damping-off

Symptoms: Post-emergenic damping-off of seeding occurs

Causal Organism: *Pythium* sp

Survival: Infected crop debris

Management

i) Field sanitation

ii) Drenching of seed bed with Mancozeb @ 3 gm/litre

iii) Deep summer ploughing. Removal and destruction of diseased plant debris

2) Leaf spot

Symptoms: Blightening of leaves occur

Causal Organism: *Septoria chrysanthemella*

***S*urvival:** Infected crop debris

Management

i) Field sanitation

ii) Spray the crop with Mancozeb @ gm/litre

iii) Deep summer ploughing. Removal and destruction of diseased plant debris

3. Grey mold

Symptoms: Greyish mouldy growth develop on flowers and stalks causing drying of flowers and stalks

Caused Organism: *Botrylits cinerea*

Managment

i) Field sanitation

ii) Spray the crop with Carbendazim 50 wp @ 1gm/litre

Minor diseases

Fungal diseases

Rust: *Puccinia chrysanthemi*

Leaf spot: *Alternaria tenuis*

Leaf blight: *Alternaria alternata*

8

Diseases of Fodder Crops

BERSEEM CLOVER (*Trifolium alexandrinum*)

Fungal diseases

i) Damping-off

Symptoms

Pre-emergence damping-off: Seeds decay inside the soil.

Post-emergence damping-off: Sometimes after germination water-soaked blighted seedlings develop. Seedlings collapse and wilt. Stem and root tissues rot below the soil surface.

Causal organism

i) *Pythium spinosum* var. *spinosum*

ii) *Rhizoctonia solani*

Favourable condition

i) Promoted by high humidity

ii) Poor air circulation.

Management

i) Field sanitation

ii) Crop rotation with non-host crops

iii) Deep ploughing in summer

iv) Use of certified seeds

v) Seed treatment with Carbendazim 50%WP (0.15%)

vi) Removal of infected plant and destruction of diseased plant debris

vii) Basal drenching with Metalaxyl @0.2% twice or thrice at 10 days interval if necessary.

2) Root Rot

Symptoms: Wilting of plants which appeared in patches under favourable condition and white hyphae turned to grey by *R. solani*. *Fusarium* species enter the healthy roots of plants and move upward through xylem tissues block the vascular bundles.

Causal organism

i) *Fusarium* sp.

ii) *Rhizoctonia solani*

Management

i) Field sanitation

ii) Crop rotation with non-host crops for three years or fallowing the field.

iii) Deep ploughing in summer

iv) Seed treatment with Carbendazim 50%WP @ (0.15%)

v) Removal of infected plant and destruction of diseased plant debris

vi) Basal drenching with Carbendazim @ 0.2% twice or thrice at 10 days interval if necessary.

3) Stem and Crown rot

Symptoms: Water-soaked lesions appear around stem, later turn to patches in severe conditions and ultimately, crop fails to grow. Sclerotia develop on infected tissues and plant debris.

Causal organism: *Sclerotinia trifoliorum*

Management

i) Use healthy plants from disease-free fields.

ii) Pre-sowing seed treatment with Thiophanate methyl @ 2–3 gram/kg

iii) Avoid over irrigation.

iv) Basal drenching with Carbendazim @ 0.2% twice or thrice at 10 days interval if necessary.

4) Stemphylum leaf spot

Symptoms: Brown to black lesions sometimes surrounded by yellow margins which enlarge and finally leaves wither from tips. Dark brown lesions are also produced on the stem.

Causal organism: *Stemphylium globuliferum*

Management

i) Field sanitation

ii) Crop rotation with non-host crops

iii) Seed treatment with Carbendazim 50% WP (0.15%)

iv) Removal of infected plant and destruction of diseased plant debris

v) Basal drenching with Carbendazim 50% WP (0.15%) twice or thrice at 10 days interval if necessary.

vi) Spraying of Mancozeb 50% WP (0.3%) /Tebuconazole 25% WG(0.1%)/ Difenconazole 25% EC(0.1%) or Hexaconazole 5% EC (0.1%) at 7 days interval

5) Curvularia leaf spot

Symptoms: Brown coloured spots appear on leaves which merged and produced black burning on leaves.

Causal organism: *Curvularia lunata*

Management

i) Field sanitation

ii) Crop rotation with non-host crops

iii) Seed treatment with Carbendazim 50% WP @ 0.15%.

iii) Removal of infected plant and destruction of diseased plant debris

iv) Spraying of Mancozeb 50%WP (0.3%) or Tebuconazole 25% WG (0.1%) or Difenconazole 25% EC (0.1%) or Hexacoazole 5% EC(0.1%) at 7 days interval

Viral diseases

1) Mosaic

Symptoms: Light green mottling which appears along veins and veinlets after each cutting. Chlorotic patches appear when the season warms.

Causal organism: Alfa alfa mosaic virus

Management

i) Field sanitation

ii) Crop rotation with non-host crops

iii) Seed treatment with Carbendazim 50% WP (0.15%)

iv) Removal of infected plant and destruction of diseased plant debris Spraying of insecticides i.e. Imidacloprid (0.05%) or Thiamethoxam (0.2%)

v) Select healthy planting material

Non-parasitic diseases

1) Potassium deficiency

Symptoms: Small reddish-brown spots on the upper surface of older leaf edges which enlarges till distal part of margin (brown scorch)

Management

i)Application of organic mulch and potassium silicate or sulfur or polymer-coated potassium products.

2) Manganese deficiency

Symptoms: Occurs in coarse texture soils where berseem follows rice.

Management

i) Most prominent at harvesting stage.

ii) Grey to yellow mottling of leaves.

iii) Spray 1kg of Mg Sulphate in 200 litre of water/acre at weekly intervals.

Minor disease

Fungal disease

Charcoal rot: *Pellicularia filamentosa*

Brown spot: *Physoderma trifolii*

Leaf spot: *Colletotrichum* sp

Root and stem rot: *Rhizoctonia solani*

Stem rot: *Sclerotinia sclerotiorum*

RED CLOVER (*Trifolium pratense*)

Fungal diseases

i) Root rot

Symptoms: Root portions decay. Cotyledons show yellowing. The plant become stunted followed by death.

Causal organism: *Aphanomyces* spp.

Management

i) Field sanitation

ii) Crop rotation for three years

iii) Deep summer ploughing. Removal and destruction of diseased plant debris

iv) Timely planting

v) Treatment of planting materials with *Trichoderma* sp @4g/kg.

2) Crown and stem rot

Symptoms: Stem portions at the ground level become covered with sclerotia show rotting. Soft rot occur on crowns followed by wilting of the plant.

Causal organism: *Sclerotinia trifoliorum*

Management

i) Field sanitation

ii) Crop rotation for three years

iii) Deep summer ploughing. Removal and destruction of diseased plant debris

iv) Timely planting

v) Treatment of planting materials with *Trichoderma* sp @ 4g/kg.

3) Powdery mildew

Symptoms: Powdery patches appear on the leaves and other green parts. Later on these patches become dull coloured and are covered with black dot like cleistothecia. The leaves become dry and fall down.

Causal organism: *Erysiphe polygoni*

Survival and spread

Primary infection is through the resting spores and secondary spread occurs through infected planting materials and wind.

Management

i) Field sanitation

ii) Crop rotation

iii) Spraying of crop with Sulphur 80% WP @ 3-4gms/litre two to three times at 15 days interval

iv) Application of judicious dose of nitrogen and phosphorus fertilizers

v) Growing of resistant or tolerant varieties

4) Black patch

Symptoms: Brown to black lesions appear on leaves

Causal organism: *Rhizoctonia legumicola*

Management

i) Field sanitation

ii) Crop rotation for three years

iii) Deep summer ploughing. Removal and destruction of diseased plant debris

iv) Timely planting

Bacterial diseases

1) Bacterial leaf spot

Symptoms: Water-soaked spots appear on the leaves. These spots enlarge and coalesce cause blightening of leaves.

Causal organism: *Pseudomonas syringae*

Management

i) Field sanitation

ii) Deep summer ploughing. Removal and destruction of diseased plant debris

iii) Timely planting

Viral diseases

1) Vein mosaic disease

Symptoms: The leaf veins become yellow and adjacent tissues turn white. Ultimately the yield reduce

Causal organism: Red clover mosaic virus

Management

i) Field sanitation

ii) Deep summer ploughing. Removal and destruction of diseased plant debris

iii) Timely planting

2) Mottling

Symptoms: The leaves show interveinal yellowing followed by chlorosis then become mottled and crinkled.

Causal organism: Red clover mottle virus

Management

i) Field sanitation

ii) Deep summer ploughing. Removal and destruction of diseased plant debris

iii) Timely planting

iv) Spraying of insecticides to control vectors

Phytoplasma diseases

1) Phyllody

Symptoms: Leaf size reduced followed by mild chlorosis. Older leaves become bronze colour.

Causal organism: Phytoplasma

Management

i) Field sanitation

ii) Deep summer ploughing. Removal and destruction of diseased plant debris

iii) Timely planting

iv) Spraying of insecticides to control vectors

2) Proliferation

Symptoms: Flowers transform into leafy like structure.

Causal organism: Phytoplasma

Management

i) Field sanitation

ii) Deep summer ploughing. Removal and destruction of diseased plant debris

iii) Timely planting

iv) Spraying of insecticides to control vectors

3) Yellow edge

Symptoms: The plants become stunted the leaves show severe chlorosis along the leaf margins and become reddish brown. There is reduction in flower size and also production

Causal organism: Phytoplasma

Management

i) Field sanitation

ii) Deep summer ploughing. Removal and destruction of diseased plant debris

iii) Timely planting

v) Spraying of insecticides to control vectors

LUCERNE (*Medicago sativa*)

Fungal diseases

1) Leaf spot

Symptoms: Round dark spots appear on upper surface of the leaves. Infected leaves turn yellow and brown. On stem brown to black elongated lesions are usually less than 2 mm in diameter appear. The leaves turn yellow and fall down. In severe infection the plants can be defoliated. There is reduction in hay quality and yield.

Causal organism: *Stemphylium botryosum*

Survival: The pathogens survive in infected crop debris. Secondary spread occurs by wind.

Management

i) Field sanitation

ii) Deep summer ploughing. Removal and destruction of diseased plant debris

iii) Timely planting

iv) Treatment of planting materials with *Trichoderma* sp @ 4g/kg

v) Spraying of crop with Chlorothalonil (0.2%) twice/thrice at 7 days interval after appearance of the disease in field.

vi) Hay crops should be harvested as early as possible to reduce crop losses.

2) Rust

Symptoms: Brown pustules appear on the under side of leaves. On the upper surface of the leaves dark-brown patches occur which is often surrounded by a welldefined chlorotic ring. In severe infection the leaves become yellow and fall. On the stem elongated pustules develop.

Causal organism: *Uromyces striatus*

Management

i) Clean cultivation

ii) Timely cutting reduces leaf loss and the build up of inoculums

iii) Spraying of Mancozeb 50%WP @0.3%

3) Spring Black stem

Symptoms: Dark brown small spots appear on the lower leaves. In severe infection the leaves become yellow, wither and fall off. Near base of the plant lesions appear. Severe infestations may girdle and kill the stem

Causal organism: *Phoma medicaginis*

Survival: Infected crop debris

Management

i) Field sanitation

ii) Timely planting

iii) Cutting the plants at early stages reduce the loss

iv) Spraying of crop with Mancozeb (0.2%) twice or thrice at 7 days interval after appearance of the disease in field.

4) Crown and Stem rot

Symptoms: A white cotton moldy growth develop on the infected tissue near the soil level and cause rapid wilting. Small, black, hard sclerotia develop which can remain viable for 15 years.

Causal organism: *Sclerotinia trifoliorum*

Survival: Infected crop debris

Management

i) Field sanitation

ii) Crop rotation for three years

iii) Deep summer ploughing. Removal and destruction of diseased plant debris

iv) Timely planting

v) Treatment of planting materials with *Trichoderma* sp @4g/kg of seed.

5) Downy mildew

Symptoms: The whitish grey mold growth appear on the lower surface of the leaf. The corresponding upper surface show yellowish green blotches. Affected stems are thicker than the healthy ones with short internodes and often giving a bushy growth at their tips.

Causal organism: *Peronospora trifoliorum*

Survival: Infected crop debris

Favourable condition: Cool humid weather favours the disease.

Management

i) Field sanitation

ii) Crop rotation for three years

iii) Deep summer ploughing. Removal and destruction of diseased plant debris

iv) Timely planting and treatment of planting materials with *Trichoderma* sp @4g/kg

v) Spray the crop with Ziram @2.5gm/litre

6) Yellow leaf blotch

Symptoms: Yellow spots develop on the leaves and they gradually increase in size become irregular in shape and form greenish brown . Finally these blotches become dark brown to black. Finally the the leaves curl down and drop.

Causal organism: *Leptotrochila medicaginis*

Management

i) Field sanitation

ii) Crop rotation for three years

iii) Grow the crop in well drained soil.

iv) Selection of resistant/tolerant varieties.

v) Spraying of crop with Mancozeb 50% WS @ (0.2%) twice/thrice at 7 days interval after appearance of the disease in field.

7) Pepper leaf spot

Symptoms: Small black pepper spots appear on younger leaves. Round to oval lesions appear on petioles and stems with darker margin and faint halo.

Causal organism: *Leptosphaerulina trifolii*

Management

i) Field sanitation

ii) Crop rotation for three years

iii) Grow the crop in well drained soil.

iv) Selection of resistant or tolerant varieties.

8) Stagonospora leaf spot

Symptoms: Dark brown specks with diffuse margin appear on leaves and stems. Severe infection cause leaf blightening.

Causal organism: *Stagonospora* sp.

Management

i) Field sanitation

ii) Crop rotation for three years

iii) Grow the crop in well drained soil.

iv) Selection of resistant or tolerant varieties.

9) Anthracnose

Symptoms: The disease appears as diamond shaped lesions with white or bleached centre on leaves and also on lower portion of the stems.In later stage small black acervuli can be seen. There may be reduction in growth of the plant. Severe infection leads to death of the plants.

Causal organism: *Colletotrichum trifolii*

Epidemiology: Humid weather favours disease development. Leaf wetness directly related to disease severity. Temperature of 27°C and RH of 80% helps its spread.

Management

i) Cut and burn the infected shoots

ii) Avoid overhead watering.

iii) Use Bordeaux mixture or Mancozeb to control the disease.

v) Spraying of crop with Mancozeb 50%WS (0.2%) twice or thrice at 7 days interval after appearance of the disease in field.

10) Root and Crown rot

Symptoms: Plants wilt and die due to the destruction of the crown bud and the root portion.

Causal organism: *Cylindrocladium floridanum*

Management

i) Field sanitation

ii) Seed treatment

iii) Grow the crop in well drained soil.

iv) Selection of resistant/tolerant varieties.

11) Summer black stem and leaf spot

Symptoms: Small brown circular spots appear on both the surface of leaves and cover the entire surface. Severely infected leaves fall down. In severe infection dark brown lesions appear on stems, petioles and peduncles.

Causal organism: *Cercospora medicagensis*

Epidemiology: Humid weather favours disease development. Leaf wetness directly related to disease severity. Temp of 27°C& RH of 80% helps its spread.

Management

i) Cut and burn the infected shoots.

ii) Removal of crop debris.

iii) Avoid overhead watering.

iv) Spraying of Mancozeb 50% WS (0.3%) to control the disease.

Bacterial diseases

1) Bacterial wilt

Symptoms: The disease occurs in patches. The plants become dwarfed. Leaves are malformed and much smaller than healthy leaves.

Causal organism: *Corynebacterium insidiosum*

Survival: Infected crop debris.

Favourable condition: Continous rainfall/high rainfall and poorly drained soil increase the disease severity. In growing stage of the crop wetter environment favours the disease development.

Management

i) Field sanitation

ii) Crop rotation for three years. Avoiding of growing potato and tomato in the same site/location at least two and preferably five years.

iii) Deep summer ploughing. Removal and destruction of diseased plant debris.

iv) Growing of resistant cultivars. Harvesting of young stands first than older one if same implements are used for harvesting.

Viral diseases

1) Mosaic

Symptoms: Plants become dwarfed. Leaves and petioles are malformed. Leaves show pale green yellow patches. Yellow green mosaic pattern or bright yellow blotches appear on leaves. As the plant matures these patches cover large areas of the leaves.

Causal organism

Alfaalfa mosaic virus

Survival: i) Volunteer plants, ii) Infected crop debris, iii) Weed hosts.

Management

i) Field sanitation

ii) Crop rotation for three years

iii) Deep summer ploughing. Removal of diseased plant and destruction of diseased plant debris

iv) Use of certified seed

v) Spraying of Imidacloprid against vector

vi) Growing of aphid resistant cultivar

Minor diseases

Fungal diseases

Leaf spot: *Alternaria brassicae*

Leaf spot: *Alternaria tenuissima*

Leaf spot: *Pseudopeziza medicaginis*

Anthracnose: *Colletotrichum trifolii*

Anthracnose: *Colletotrichum medicaginis*

Cercospora leaf spot: *Cercospora medicaginis*

Cercospora leaf spot: *Cercospora medicaginicola*

Fusarium wilt: *Fusarium oxysporum* f.sp *medicaginis*

Root rot: *Phytophthora megasperma*

Charcoal rot: *Macrophomina phaseolina*

Powdery mildew: *Leveillula taurica*

GUAR (*Cyamopsis tetragonoloba*)

Alternaria leaf spot

Symptoms: Round to irregular dark brown spots appear on the leaves. Later on these spots turn greyish to dark brown. Concentric rings are present in these spots. Several spots may coalesce to form bigger necrotic patches. The leaflets become completely yellow and drop off. If the plants are infested in early stage they may not produce any flower.

Causal organism: *Alternaria cyamopsidis*

Management

i) Field sanitation.

ii) Use of certified seeds or sowing of seeds collected from healthy plants of disease free field.

iii) Seed treatment with Carboxin 37.5% WP +Thiram 37.5% WP @ 3gm/kg of seed.

iv) Spraying of crop with Mancozeb @ 0.3% after appearance of the disease in the field. If necessary two or three sprayings may be practised at 7days interval.

Anthracnose

Symptoms: Circular dark brown spots appear on the leaves. These spots increase in size become dark lesions. In moist weather black fruiting bodies appear on these lesions. In severe infection petioles and stems are infected.

Causal organism: *Colletotrichum capsici* f.sp. *cyamopsicola*

Management

i) Field sanitation.

ii) Use of certified seeds or sowing of seeds collected from healthy plants of disease free field.

iii) Seed treatment with Carboxin 37.5% WP +Thiram 37.5% WP @ 3gm/kg of seed.

iv) Spraying of crop with Mancozeb 50%WS@ 0.3% after appearance of the disease in the field. If necessary two or three sprayings may be practised at 7days interval.

Powdery mildew

Symptoms: White powdery patches appear on both the surface of leaves. In severe infection the infected leaves dry up followed by defoliation.

Causal organism: *Leveillula taurica*

Management

i) Field sanitation.

ii) Seed treatment with Carboxin 37.5% WP +Thiram 37.5% WP @ 3gm/kg of seed.

iii) Spraying of crop Sulfex / Karathane @ 0.2% after appearance of the disease in the field. If necessary two or three sprayings may be practised at 7days interval.

Bacterial diseases

Bacterial Blight

Symptoms: Interveinal round spots appear on the dorsal surface of the leaves. Gradually these spots enlarge, coalesce and cause blightening of the leaves. The pathogen cause flaccidity of infected portions. The flaccid spots become necrotic and turn brown. In severe infection longitudinal streaks appear on stems and petioles cause blackening and cracking of the stem.

Causal organism: *Xanthomonas cyamopsidis*

Management

i) Field sanitation.

ii) Use of certified seeds or sowing of seeds collected from healthy plants of disease free field.

iii) Seed treatment with Carboxin 37.5% WP +Thiram 37.5% WP @ 3gm/kg of seed.

iv) Hot water treatment of seeds at 56 ^{0}C for 10minutes.

Other minor diseases

Fungal diseases

Damping off : *Pythium myriotylum*

Root rot: *Sclerotium rolfsii*

Charcoal rot: *Macrophomina phaseolina*

White rot: *Pellicularia rolfsii*

Root rot: *Phymatotrichum omnivorum*

Root rot: *Sclerotium rolfsii*

Leaf spot and blight: *Alternaria alternata*

Leaf spot: *Alternaria brassicae*

Leaf spot: *Cercospora canescens*

Leaf spot:*Curvularia lunata*

Leaf spot: *Drechslera australiensis*

Bacterial diseases

Bacterial leaf stripe:*Pseudomonas sorghicola*

Viral diseases

Mosaic: Enation mosaic virus

Sunhemp mosaic virus

SUDAN GRASS (*Sorghum sudanensis*)

Fusarium stalk rot and top rot

Symptoms: There may be post-emergence damping-off seedlings. If the infection is late in the infected plants, the leaves can not emerge freely .They remain twisted and rolled with each other. The upper nodes become shorter and discoloured .The dark lesions may be developed on outside of the nodes. In infected stalk the pith first become water soaked and then turns deep red. The affected plants rarely flowers.

Causal organism:*Fusarium moniliforme*

Management

i) Removal of stubbles from the field after harvesting of the crop.

ii) Clean cultivation

iii) Growing of resistant varieties.

iv) Crop rotation.

v) Use of certified seeds.

vi) Seed treatment with Carboxin 37.5% WP +Thiram 37.5% WP @ 3gm/kg of seed.

vii) Spraying of crop with Mancozeb @ 0.3%

Ascochyta leaf spot

Symptoms: Small,circular to oblong small light coloured spots with well defined margins develop on the leaves. Gradually these spots enlarge in size and coalesce to form large blotches. Black raised pycnidia develop on these blothches giving them rough appearance. The lesions become yellow brown or purple red in colour. The matured leaves become necrotic..

Causal organism: *Ascochyta sorghina*

Management

i) Growing of resistant varieties.

ii) Use of certified seeds.

iii) Crop rotation

iv) Seed treatment with Carboxin 37.5% WP +Thiram 37.5% WP @ 3gm/kg of seed.

v) Spraying of crop with Mancozeb @ 0.3%

Anthracnose and red leaf spot

Symptoms: Small reddish spots appear on the leaves. The centre of the spots become whitish with reddish/purplish or brownish margin. In later stage small black dot like fruiting bodies appear. Plants show defoliation. In severe infection there may be premature death of the plants.

Causal organism: *Colletotrichum graminicolum*

Management

i) Growing of resistant varieties.

ii) Use of certified seeds.

iii) Seed treatment with Carboxin 37.5% WP +Thiram 37.5% WP @ 3gm/ kg of seed.

iv) Spraying of crop with Mancozeb @ 0.3%

Sooty Stripe

Symptoms: Small water-soaked oblong reddish purple spots appear on older leaves.Gradually these spots increase in size and develop into elongated to elliptical lesions. Initially the centres of the lesions become yellowish brown which ultimately turn necrotic with reddish purple to yellowish brown border. In later stage the centres of the lesions give sooty appearance due to formation of large number of sclerotia on them.

Causal organism: *Ascochyta sorghina*

Management

i) Removal of stubbles from the field after harvesting of the crop.

ii) Clean cultivation

iii) Growing of resistant varieties.

iv) Use of certified seeds.

v) Crop rotation.

vi) Seed treatment with Carboxin 37.5% WP +Thiram 37.5% WP @ 3gm/ kg of seed.

Ergot

Symptoms: In this disease the individual spikelets are infected. The typical honey dew secretion take place from the affected spikelts. The grains are replaced by soft mass of fungal growth.

Causal organism: *Sphacelia sorghi*

Management

i) Growing of resistant varieties.

ii) Use of certified seeds.

iii) Seed treatment with Carboxin 37.5% WP +Thiram 37.5% WP @ 3gm/ kg of seed.

iv) Spraying of crop with Mancozeb @ 0.3%

Viral diseases

Mosaic

Symptoms: Longitudinal yellow and white streaks with irregular mottling of dark and green areas are seen on the upper two to three leaves. Severe infection may cause death of the plants.

Causal organism: Maize dwarf mosaic virus

Management

i) Removal of stubbles from the field after harvesting of the crop.

ii) Clean cultivation

iii) Growing of resistant varieties.

iv) Use of certified seeds.

v) Spraying of the crop with suitable insecticides after appearance of the disease in the field to control the vectors

References

Akinyele, B.O. and Odiyi, A.C. 2007. Comparative study of vegetative morphology and the existing taxonomic status of Aloe vera L. Journal of Plant Sciences.2(5):558-563.http://www.academic journals,net/2/archive.php?id=2s/theme=2tjid=jps

Amura, C .2020.Aloe vera. Complete Guide .My Garden

Asghar, M., Baig, M.M. Q., Raza, A.M., Arshad, M., Hussian, M and Afzal, M.S. 2019. Evaluation of the effect of chemicals used as seed dressing for control of berseem root. Fuuast Journal of Biology. 9(1):47-50

Barbetti, M.J. 1983. Fungicidal control of damping off and seedling root rot in Sub terranean clover.Fungicide Nematicide Tests.38:47

Chaudhary, A.R., Haq, I. and Reman N. 1992. A cultural approach towards controlof berseem root rot.Pakistan Journal of Agricultural Sciences. 29(1):65-68

Doidge Ethel M. 1916. Notes on some lucerne diseases. The Agricultural Journal of Sabinet African. 229-232

Guide.http://mygardenguide.com/succulents/aloe-vera//#cultivation _technique

Gupta, M.L., Kumar, S., Pandey, R., Shukla, R.S., Khaliq, A., Khaliq, A., Kalra and A., Singh and H.N. 2000. Leaf blight disease and its effect on essential oil content of palmarosa. Journal of Medicinal and Aromatic Plant Science. 22(1B):504-505

Gyanendra, S. and Verma R.K. 2005. Compatibilityof fungicides and neem products and Trichoderma sp. against Fusarium solani f.sp.glycines causing root rot of soybean. Journal of Mycopathological Research. 43:211-214.

Haq, U.I.I. and Ijaz, S. Sustaiable Winter Fodder Production ,Chllanges and Prospects.CRC Press Taylor & Francis Group.

Iqbal, M.F. and Iqba, I.Z. 2014. Efficacy of Fungicides sprayed against rottening of berseem. International Journal of Advanced Multidisciplinary Research.1(2) :22-24.

Jetawat, R.P.S. and Mathur, K. 2016. Management of Ashwagandha root rot disease with fungicides,biocontrol agents and botanicals.Journal of Applied and Natural Science. 8(1):305-309.

Jobshy, Z.M, Syed, E.L., Rammah, A and Satter, M.A. 1981. Pathogenicity and control of three fungi associated with damping off and root rot of Egyptian Clover Trifolium alexandrinum. Research Bulletin.1674:14.

Karla, A., Singh, H.B., Pandey, R., Samad, A., Patra, N.K. and Kumar, S. 2008. Diseases in Mint,Causal organisms, Detection and Control measures.Journals of Herbs,Spices and Medicinal Plants.11(1-2):71-91.

Khan, I., Nawaz, M., Shahid, M., Zamir, I., Bibi, F., Mahmood, F., Ali, M., Chattha, MU Iqbal, M., Amjad,T., Hussain, S., Hanif, M.S., Hassan, M.U. and Usman, S. Bioecology and Agronomy of Winter Fodder Crops. DOI:10.1201/97810030553652

Pande, S.K., Chaubey, A.N., Singh, V. and Pal, K. 2017. Eco friendly management of cercospora leafspot (Cercospora rauwolfia) of sarpgandha.Plant Archives. 17(1):546-548https://www.researchgate.net.>publication>35110267

Rangaswami,G. and Mahadevan, A. 2005. Diseases of crop plants in India, Fourth Edition, 536 Pp, Prentice-Hall of India Co.Pvt Limited Publishers,New-Delhi-110001.

Saha, L.R., 2002. Handbook of plant diseases, 443Pp, Kalyani Publishers, New Delhi-110002

Shukla R.S., Khaliq A. and Alam, M. 2010. Chemical control of blossom blight disease of sarpagandha caused by Colletotrichum capsici. African. Journal of Biotechnology. 38:6397-6400

Siddra, Haq UI.I., Habib, Z., Mukhtar, S. and Nasir, B. Biotechnological applications for developing resistance and other quality tracts in fodder crops.

Singh, R.P., Singh, A.K., Singh, M. and Singh, R.K. 2020. Diseases in berseem and its management:A review.Journal of Pharmacognoy and Phytochemistry.9(3) :2054-2057.

Singh, R.S. 2001, Plant Diseases, Seventh Edition,686Pp, Oxford &IBH Publishing Co.Pvt-Ltd, New Delhi-110001

Thakur, R.N.1994.Curvulariablight-anewdiseaseofpalmarosa.Indianphytopathology.47(2):214

Thygarthi R, Parahuram, T.R., Byrappa, S.M and Manchanahally, B.S. 2013. Fungal foliar diseases of Rauwolfia serpentina in wild, its seasonal occurrence, seed transmission and disease management. Archives of Phytopathology and Plant Protection. 46(13)

Vyas, S.S. (1994). Integrated biological and chemical control of dry rot on soybean.Indian Journal.Mycol.Plant Pathol.24:132-134.